Introduction to
Agricultural Accounting

Introduction to
Agricultural
Accounting

Barbara M. Wheeling

Montana State University—Billings
Billings, Montana

THOMSON
™
DELMAR LEARNING

Australia Canada Mexico Singapore Spain United Kingdom United States

Introduction to Agricultural Accounting
Barbara M. Wheeling

Vice President, Career Education Strategic Business Unit:
Dawn Gerrain

Director of Learning Solutions:
John Fedor

Acquisitions Editor:
David Rosenbaum

Managing Editor:
Robert L. Serenka, Jr.

Product Manager:
Christina Gifford

Editorial Assistant:
Scott Royael

Director of Production:
Wendy A. Troeger

Production Manager:
Mark Bernard

Senior Content Project Manager:
Dianne Toop

Technology Project Manager:
Sandy Charette

Production Assistant:
Matthew McGuire

Director of Marketing:
Wendy E. Mapstone

Marketing Manager:
Gerard McAvey

Marketing Coordinator:
Jonathan Sheehan

Art Director:
Joy Kocsis

Cover Design:
Studio Montage

Cover Images:
Getty Images

For permission to use material from this text or product, submit a request online at http://www.thomsonrights.com

Any additional questions about permissions can be submitted by email to thomsonrights@thomson.com

Library of Congress Cataloging-in-Publication Data

Wheeling, Barbara M.
   Introduction to agricultural accounting/Barbara M. Wheeling.
      p. cm.
   Includes bibliographical references and index.
   ISBN-13: 978-1-4180-3834-2 (alk. paper)
   ISBN-10: 1-4180-3834-2 (alk. paper)
   1.   Agriculture—Accounting. I. Title.
   HF5686.A36W44 2007
   657'.863—dc22

2007018257

NOTICE TO THE READER

Publisher does not warrant or guarantee any of the products described herein or perform any independent analysis in connection with any of the product information contained herein. Publisher does not assume, and expressly disclaims, any obligation to obtain and include information other than that provided to it by the manufacturer.

The reader is expressly warned to consider and adopt all safety precautions that might be indicated by the activities herein and to avoid all potential hazards. By following the instructions contained herein, the reader willingly assumes all risks in connection with such instructions.

The Publisher makes no representation or warranties of any kind, including but not limited to, the warranties of fitness for particular purpose or merchantability, nor are any such representations implied with respect to the material set forth herein, and the publisher takes no responsibility with respect to such material. The Publisher shall not be liable for any special, consequential, or exemplary damages resulting, in whole or part, from the readers' use of, or reliance upon, this material.

DEDICATION

To my parents and grandparents
The first farmers that I ever knew

CONTENTS

PREFACE

In this first edition, *Introduction to Agricultural Accounting* explains and applies basic accounting principles to agricultural production. The changes in the agricultural industry since the early 1980s have increased the need for more complete and accurate farm financial information. Many short-term and long-term decisions in production agriculture require techniques for analyzing the financial position and financial performance of the farm operation.

Several other textbooks cover the wide range of topics necessary for students in agriculture to learn about farm management. Some of these books assume that students have studied accounting in other courses and offer an overview of the application of accounting principles to agricultural operations. In many agricultural curriculums, students take accounting courses offered in business departments at their college or university. Unfortunately, these courses do not address some of the issues and situations unique to the agricultural industry. This book offers an approach for compiling and analyzing accounting information that is more relevant and applicable for agricultural operations than is offered in the typical business accounting course.

This book is written for students in colleges and universities with agricultural programs and for farmers and ranchers, agricultural lenders, and other advisors. It can be used as a beginning text in accounting and is designed for undergraduate students and others without any accounting background. The accounting concepts in this book are based on generally accepted accounting principles used in the United States and the *Financial Guidelines of Agricultural Producers* developed by the Farm Financial Standards Council. This book is a starting point for understanding and applying these principles and guidelines. This book can also be studied by those with an accounting background. Those readers who know basic bookkeeping procedures may want to skim through Chapters 2 and 3, which explain accounts and recording transactions. However, the remaining chapters explain those techniques that are unique to farm accounting.

The main focus of the book is the preparation and analysis of financial statements. Budgeting and other managerial accounting topics are left for other books. Double-entry accounting is emphasized. Cash-basis accounting is prevalent in many farm operations and forms the basis for the accrual-adjusted system that is explained in this book. The accrual-adjusted system provides more accurate information for the measurement of profitability of the farm operation. Market-based financial statements are also discussed as relevant information for measuring the farm financial position.

The first five chapters describe the purpose and formats of farm financial statements, define the types of accounts commonly used in agricultural operations, and explain how to record transactions and prepare financial statements under the accrual-adjusted system. Chapters 6, 7, 8, and 9 specify how to measure the various items found in farm financial statements. Chapter 10 describes the various methods of analyzing financial position and financial performance.

The book is written with the assumption that instructors, students, and other readers will follow the chapter sequence as presented. The last chapter on financial ratios and other analytical tools sometimes draws the greatest interest. Many people may wish to know only how to analyze the farm's financial performance. For that reason, some instructors may wish to begin the course by introducing these analytical techniques and then proceed to instruct students on the techniques for compiling meaningful accounting information so that the financial ratios also have meaning. Another advantage for this approach is that students realize where they are being led so that they know the answer to the question of "why do we need to know this?" Whether the instructor begins or ends the course of study with Chapter 10, Chapters 1 through 9 should be studied in sequence because examples and concepts in later chapters build upon the earlier chapters.

The original idea for the book came to me after a conversation with Richard Benson, a farmer in Wisconsin and a former beef cattle extension specialist, who told me about the work of the Farm Financial Standards Council. At that time, I was teaching accounting at Southern Utah University to students who included agribusiness majors. I realized that some of the accounting subject matter did not apply to agriculture and, suddenly, I had the idea to write a book. Benson has been a strong supporter of this endeavor, offering me much encouragement, which I needed especially during the time that I was having difficulty finding a publisher. Thank you, for your support.

I would also like to thank the Farm Financial Standards Council for their work. It took more than a decade to develop much needed guidelines for farmers, ranchers, agricultural students, lenders, and others in the agricultural industry. When I was involved in agriculture, I wished that I had known more about accounting. It was not until years later that I received my education. The *Financial Guidelines for Agricultural Producers* published by the Council brings accounting to agriculture in more detail than other professional pronouncements. I would like to thank members of the Council who have expressed interest and support of this book, particularly John Larson and Joe Daughettee, for their helpful comments on an early draft of this book and their continued support for completing this project.

My thanks also extend to the reviewers of the proposal and the first draft of the book. Although they remain anonymous, I am grateful for their help in developing my early draft into a publishable manuscript. I hope that I have addressed their ideas adequately. I would also like to thank a former student, Jodi Tomky, who gave me early feedback on the manuscript. It was encouraging to me when she said she wished that someone had taught her the basics of accounting in the way that I had presented it. I also thank Jennifer Yohe, a friend who read the early manuscript and taught me a few things about editing. Thank you, all.

I am also very grateful to David Rosenbaum, the acquisitions editor at Thomson Delmar Learning, who was interested in acquiring the book for publication. After searching a long time for a publisher, I was greatly relieved and appreciative of his interest and enthusiasm toward this project. My thanks extend to all at Delmar who have worked on this book, especially to Christina Gifford, who has kept me on track as my main liaison with Delmar. This book would not have been published without your help. Thank you.

Finally, I would like to thank the family members, friends, and colleagues who have encouraged me during the work on this book. They are too numerous to mention, but I would like to mention my sister, Yvonne, who is a certified public accountant with many agricultural clients, for her interest in this subject, my supportive colleagues at Montana State University in Billings, Montana, and John

Staffeld, a former colleague who reviewed my publishing contract and has been excited about my getting this book published from the very beginning. Thank you all for your support. Your encouragement helped me to persevere.

I also have some very close friends whose support I can always count on. I will thank you all in person when we celebrate this book.

To all those who read this book, I hope that you are able to learn something from it. I may be reached by e-mail at bwheeling@msubillings.edu or by phone 406-657-1756.

Barbara M. Wheeling

Barbara M. Wheeling received her Ph.D. in Accounting from the University of Alberta in 1999. She also has a Masters degree in Business Administration with an Accounting and Finance emphasis from the University of Wyoming (1989) and a Bachelor of Science degree in Animal Science from North Dakota State University (1972). She is currently serving as assistant professor in accounting at the College of Business, Montana State University—Billings, and has also served on the faculties of Colorado State University—Pueblo and Southern Utah University. Her publications include "Alternatives to GAAP for the Agricultural Industry," published in *Today's CPA* (November/December, 2005), "Agricultural Accounting: A Case for Asset Revaluation," published in the *Mountain Plains Journal of Business and Economics* (Vol. 2, 2001), and *Accounting for Agricultural Producers*, a monograph for the Bureau of National Affairs, Washington, D.C. (2007). She is a member of the Northwest Farm Business Management Instructors and attends meetings of the Farm Financial Standards Council.

CHAPTER 1

Farm Financial Statements

Key Terms

Accounting
Accounting Equation
Accrual Adjusted Net Farm Income
Accrual Adjusted Net Income
Assets
Balance Sheet
Change in Excess of Market Value over Cost/Tax Basis of Farm Capital Assets
Change in Non-Current Portion of Deferred Taxes
Current Assets
Current Liabilities
Depreciation Expense

Disclosure by Notes
Equity
Expenses
Farm Financial Standards Council
Farm Financial Statements
Farm Revenues
Financial accounting
Financial Accounting Standards Board
Financial analysis
Financial performance
Financial position
Financial statements
Financing activities
Gains

Generally Accepted Accounting Principles
Gross Revenue
Income before Taxes
Income Statement
Income Tax Expense
Interest Expense
Investing activities
Investments
Liabilities
Losses
Net Farm Income from Operations
Net Income or Net Profit
Non-Current Assets

Non-Current Liabilities
Non-Farm Income
Operating activities
Operating Expenses
Operating Income
Other Capital Contributions/Gifts/ Inheritances
Other Expenses
Other Revenues
Retained Capital
Revenues
Statement of Cash Flows
Statement of Owner Equity
Valuation Equity
Withdrawals

Farming is a business. Ills of many industrial and commercial businesses in our large cities are corrected by the inauguration of better financial, accounting and marketing policies. . . . Many of the ills of the farming business likewise can be cured by good management from within rather than legislation from without the farm boundaries. Accounting is one of the best tools of the successful manager anywhere. If the farmer can keep accounts even on a simple basis and learn to use them properly he will be in a better position to detect his weaknesses in producing and marketing the products of the farm.

So wrote Hiram T. Scovill in his review of *Farm Accounting Principles and Problems*, a book by Karl F. McMurry and Preston E. McNall. The book was published in 1926 and the review appeared shortly thereafter in *The Accounting Review*, an accounting academic journal that still publishes research papers in accounting. Professor Scovill's words are no less relevant today than they were in the 1920s. Many technological, environmental, political, financial, social, and global changes have occurred since then, but the need for good management (including accounting) remains. Because changes occur rapidly, good management and good accounting in agriculture may be more important than ever before.

This book addresses one aspect of farm management—the tool that Professor Scovill refers to as one of the best that a successful manager must use—accounting. Financial management involves the acquisition and use of capital, the minimization of risk, financial planning and analysis, and relationships with financial institutions, to name a few. Accounting helps provide information for making decisions. Good decision making and good management are hampered by incomplete, inaccurate, or inconsistent information. The purpose of this book is to help you, the reader, to learn and apply accounting principles and procedures to help develop good farm management practices.

In this chapter, you will learn why financial information about a farm business is required. At the end, you will be able to define "financial statements" and describe the characteristics of these statements that make them useful to others. You will also be able to define "financial accounting" and to identify the rules for financial accounting and the parties that developed them. You will also be able to define terms including financial analysis, financial position, and financial performance. You will understand the meaning of the term "capital," describe the three types of financial activities, and provide some examples of each activity. You will be able to name the four primary Farm Financial Statements, describe the general purpose or function of each of them, and name the elements contained in each statement. You will learn how to define and calculate "net income" or "net loss," state the general format for the Income Statement, and define or describe each of the items or categories in it. You will also be able to explain the advantage of the expanded format of the Income Statement, and state the main differences between the general format and the expanded format. You will learn how to define the concept of "equity," describe the general and expanded formats for the Statement of Owner Equity, and define or describe each of the items or categories in it. You will also be able to state the Accounting Equation and the general format for the Balance Sheet and to define or describe the items or categories in it. You will be able to identify the components of equity, name the three categories of activities reported on the Statement of Cash Flows, provide examples of each category, and outline the general format for the Statement of Cash Flows.

OVERVIEW OF AGRICULTURAL ACCOUNTING

Learning Objective 1 ▉ To explain why financial information about a farm business is required.

Accounting is an information system that provides financial information about an economic entity. Economic entities include farms and ranches, and other types of businesses. Financial information includes many types of reports and documents with financial data. Farm and ranch managers and owners use these reports to evaluate the success or failure of the farm business and to help decide future activities. Outside parties, such as agricultural lenders, who must also make decisions in conducting business with agricultural operations, request some of these reports.

Learning Objective 2 ▉ To define "financial statements" and describe the characteristics of these statements that make them useful to those who read them.

This book is primarily about reports submitted to outside parties. These reports are called **financial statements**. Financial statements summarize the results of the financial activities of a farm operation. These statements present this information in terms of dollar value.[1]

---

1. Financial Accounting Standards Board. "Recognition and Measurement in Financial Statements," *Statement of Financial Accounting Concepts No. 5* (Stamford, CT: 1984).

You will learn that financial statements summarize activities for an entire year or for part of a year.[2] Farm managers and owners and outside parties need financial information on a periodic and timely basis,[3] so financial statements should be prepared at least once a year. They can gain useful information from financial statements in evaluating their farm activities from one year to the next. The primary reports used by farm managers, owners, and outside parties are the four types of financial statements discussed in this chapter. These four statements are the Income Statement, the Statement of Owner Equity, the Balance Sheet, and the Statement of Cash Flows.

For financial statements to be useful to farm managers, owners, and outside parties, the accounting system must provide understandable information.[4] One way to accomplish this is for each type of statement to follow a format that everyone recognizes. This chapter presents a general format and specific examples of detailed formats for each of the four financial statements. Whether you are preparing the financial statements or using them for some type of analysis, you cannot understand them without knowing the rules and guidelines used to produce the information. These rules and guidelines are developed so that the financial statements are not biased, a concept called neutrality.[5] To minimize bias, any estimates should be as conservative as possible. Furthermore, for financial information to be useful, the rules and guidelines must be used consistently, so that owners and managers can evaluate the performance of the farm business from one year to the next.[6] Lenders and other outside parties may want to compare one farm to another. That task is easier if these rules and guidelines are followed consistently by the farms they are comparing.[7]

Information summarized in the statements must be reliable for financial statements to be useful.[8] The system in this book helps to make financial statements reliable because it requires a balance in the financial numbers. For the information to be reliable, it should be verified by documents from which the information originated,[9] or source documents. They should be stored and organized

Presenting financial information in terms of dollar value might seem like an obvious statement, except when you realize that in the 19th century, farmers presented balance sheets in terms of the number of acres, pounds, bushels, head of livestock, and so on. Agricultural producers understand the need for production records that use these modes of measurements. Sometimes the dollar value is a difficult measure, for reasons that are clarified throughout this book.

---

2. Ibid.
3. Financial Accounting Standards Board. "Qualitative Characteristics of Accounting Information," *Statement of Financial Accounting Concepts No. 2* (Stamford, CT: May, 1980).
4. Ibid.
5. Ibid.
6. Ibid.
7. Ibid.
8. Ibid.
9. Ibid.

for easy reference. This requirement to verify information results in reporting many financial statement items at the original (historical) cost of the item.[10] Historical information contributes to unbiased information. To enhance reliability, the information should also be as accurate and complete as possible. Furthermore, titles identify financial statement items. It is important to learn these titles, to understand what the titles describe.

For financial statements to be useful, they must be relevant.[11] You will learn about an approach that helps to make financial statements relevant because it helps to provide complete and up-to-date information. Relevant financial statements provide feedback on past activities of the farm operation and some ability to predict future financial activities on a timely basis.[12] Relevant financial statements provide information that is likely to make a difference in decisions. The farm accountant has to be aware of the benefits that the information provides. If the information is insignificant for making decisions (a concept called "materiality"), then it should not be reported.[13] The usefulness of the information has to exceed the cost of recording and reporting it.[14]

The rules and guidelines for accounting have been developed with several other concepts in mind. One of these is that the farm business is a separate entity from the owners.[15] You will learn that Farm Financial Statements should be prepared without any personal items reported. Some of the accounting procedures reflect the principle that the farm business will continue to operate indefinitely, a concept called "going concern."[16] A concept called "revenue recognition" forms the basis for when financial activities should be reported.[17] This concept goes along with an idea called matching, which designates that financial transactions should be reported when they occur, even if the money has not been paid or received.[18] Finally, a concept called "full disclosure" requires that financial statements be accompanied by additional information that clarifies the numbers in the financial statements and the methods used to calculate them.[19]

The **Financial Accounting Standards Board (FASB)** and other groups have developed financial accounting rules or procedures (also called "standards") for companies in the United States. These accounting standards, called **Generally Accepted Accounting Principles (GAAP)**, provide a standardized system

Steve and Chris Farmer started their farming operation in the year 20X1. At the end of their first farming year, they found themselves short of cash, so they arranged to meet with their agricultural lender to negotiate for an operating loan until they could generate some positive cash flow. Their lender agreed to meet with them if they brought financial statements to the meeting. Appendix A presents the financial statements that they took to the meeting. The statements are incomplete (they do not include all of the revenues and expenses for an entire year) but as you study this book, you will learn about how they prepared these statements, and what they did to help ensure that the financial statements are useful for themselves and the lender.

---

10. Financial Accounting Standards Board. "Recognition and Measurement in Financial Statements," *Statement of Financial Accounting Concepts No. 5* (Stamford, CT: 1984).
11. Financial Accounting Standards Board. "Qualitative Characteristics of Accounting Information," *Statement of Financial Accounting Concepts No. 2* (Stamford, CT: May, 1980).
12. Ibid.
13. Financial Accounting Standards Board. "Recognition and Measurement in Financial Statements," *Statement of Financial Accounting Concepts No. 5* (Stamford, CT: 1984).
14. Ibid.
15. Ibid.
16. Ibid.
17. Ibid.
18. Ibid.
19. Ibid.

for financial accounting. **Financial accounting** refers to the accounting systems that produce financial statements for farm managers and owners and for outside parties. As indicated earlier, financial statements provide a summary of the financial activities of the farm operation. Farm managers and owners can also benefit from other accounting systems that produce more details about the farm activities, for example, the cost of production. These systems are usually designated as management (or managerial) accounting systems, which are not discussed in this book.

Standardized systems of accounting are useful for outside parties to compare more than one business. People who wish to invest in or lend money to a company need to know that similar procedures are used to summarize financial information. If that were not the case, it would be difficult to know which company is performing the best. Outside parties that are interested in doing business with an agricultural operation (such as lending money) sometimes also need to evaluate the performance of a farm operation and to compare one farm operation to others to make the best decisions about working with them.

Many agricultural operations are small, family-owned farms with simple and incomplete financial information. GAAP is often not relevant for these agricultural operations. A group called the **Farm Financial Standards Council (FFSC)** developed and published a set of recommendations in an attempt to standardize accounting procedures for agricultural operations. Their recommendations for farm and ranch accounting are called *Financial Guidelines for Agricultural Producers* (the Guidelines). The FFSC recommends the use of GAAP in preparing financial statements for agricultural operations, for the most part. However, the FFSC recognizes that many farm businesses cannot generate the necessary information to conform to GAAP completely. Therefore, the FFSC has identified alternatives to GAAP that the FFSC believes are adequate for providing useful information.

GAAP and the FFSC Guidelines provide the basis for the accounting principles and procedures that you are learning about in this book. Figure 1-1 portrays the relationship between the sources of accounting rules and their usefulness. When you have studied this book, you will have a basic understanding of accounting and how it is applied in an agricultural operation. Each farm business has unique characteristics and situations that may require complex accounting procedures. More information is available on the FFSC Website and from certified public accountants and farm consultants for these situations. The remainder of this chapter discusses the financial activities of agricultural operations and presents an overview of Farm Financial Statements.

Learning Objective 3 ■ To define "financial accounting" and to identify the rules for financial accounting and the parties that developed them.

FARM FINANCIAL ACTIVITIES AND FINANCIAL STATEMENTS

Farm Financial Statements report financial accounting information about an agricultural operation in a specified format. Financial accounting refers to the accounting system that prepares these financial statements. Financial reporting entails providing the financial statements along with additional details (the full disclosure concept). This additional information is reported in **Disclosure by Notes** (or Notes to the Financial Statements) and often includes schedules and explanations of computations. **Financial analysis** is a set of procedures in which a farm's financial position and financial performance is evaluated. **Financial position** refers to the farm's financial state at a specified point in time. **Financial performance** measures how profitable the farm was over a specified period, usually a year.

Learning Objective 4 ■ To define the following terms: "financial analysis," "financial position," and "financial performance."

FIGURE 1-1 ■ *Sources and Uses of Accounting Information.*

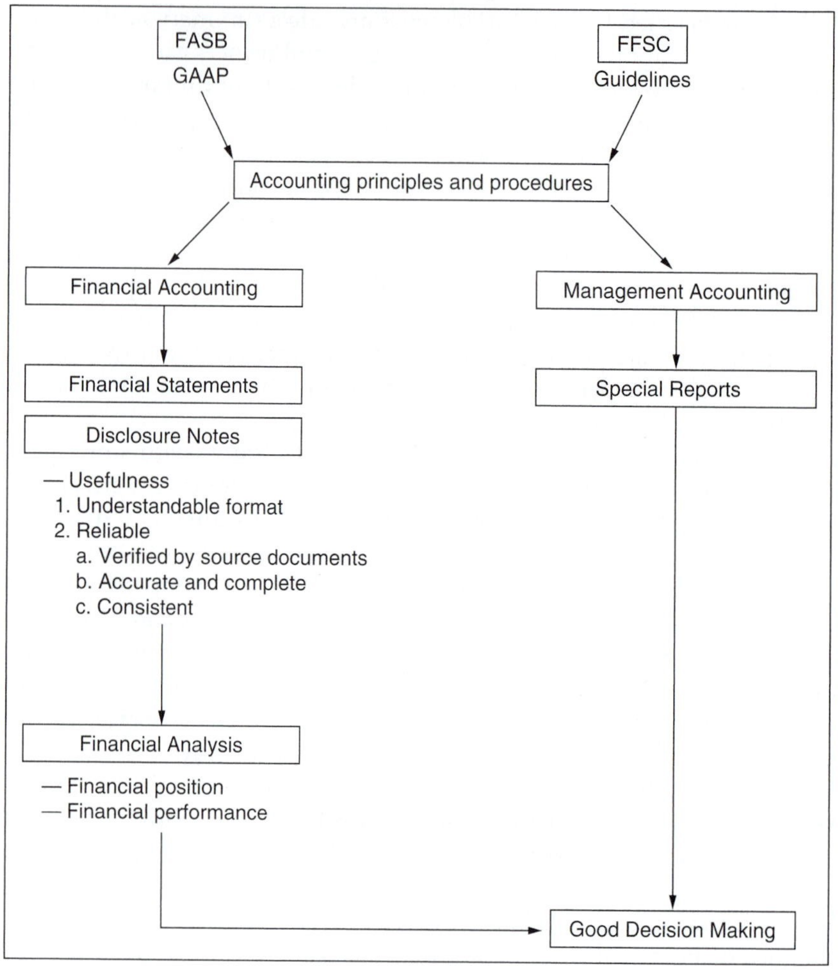

```
┌──────────────────────────────────────────────────────────────────────────┐
│        ┌──────────┐                              ┌──────────┐              │
│        │   FASB   │                              │   FFSC   │              │
│        └──────────┘                              └──────────┘              │
│          GAAP                                    Guidelines                │
│               ↘                                    ↙                       │
│              ┌─────────────────────────────────────────┐                  │
│              │   Accounting principles and procedures   │                  │
│              └─────────────────────────────────────────┘                  │
│               ↙                                    ↘                       │
│   ┌──────────────────────┐            ┌──────────────────────┐            │
│   │ Financial Accounting  │            │ Management Accounting │            │
│   └──────────────────────┘            └──────────────────────┘            │
│             ↓                                    ↓                         │
│   ┌──────────────────────┐            ┌──────────────────────┐            │
│   │  Financial Statements │            │    Special Reports    │            │
│   └──────────────────────┘            └──────────────────────┘            │
│   ┌──────────────────────┐                                                 │
│   │   Disclosure Notes    │                                                 │
│   └──────────────────────┘                                                 │
│     — Usefulness                                                           │
│     1. Understandable format                                              │
│     2. Reliable                                                           │
│        a. Verified by source documents                                    │
│        b. Accurate and complete                                           │
│        c. Consistent                                                      │
│             ↓                                                             │
│   ┌──────────────────────┐                                                 │
│   │   Financial Analysis  │                                                 │
│   └──────────────────────┘                                                 │
│     — Financial position                                                  │
│     — Financial performance                                               │
│                                                    ↓                       │
│                                     ┌──────────────────────┐              │
│             └──────────────────────→│  Good Decision Making │              │
│                                     └──────────────────────┘              │
└──────────────────────────────────────────────────────────────────────────┘
```

The items for the financial analysis are reported on the financial statements. Chapter 10 explains the procedures for financial analysis in more detail.

The first step in developing and using a farm financial accounting system is to understand which financial statements need to be prepared by the farm owner, manager, or accountant, what kind of information is to be included in each of the statements, and the format for each of the statements. With this understanding, the farm owner or manager knows what to provide for outside parties and for internal purposes. This chapter explains these concepts first by relating financial activities to the four Farm Financial Statements, and then describing the purpose and format of each of the statements.

PRACTICE WHAT YOU HAVE LEARNED   *At this point, you should be able to complete Problem 1-1 at the end of the chapter.*

Learning Objective 5 ■ To define "capital," to describe the three types of financial activities, and to provide some examples of each activity.

An agricultural operation engages in three types of financial activities: financing, investing, and operating. In **financing activities**, capital is provided to the

TABLE 1-1 ■ *Types and examples of farm financial activities.*

Financing Activities	Investing Activities	Operating Activities
Capital contributions	Purchase of assets	Production/sale of farm products
Gifts and inheritances	Sale of assets	Purchase of inventory
—received		Purchase of supplies
—distributed to heirs		Purchase of services
Non-Farm income contributed		Payment of wages
Withdrawals by owners		Payment of interest
Borrowing money		Payment of taxes
Paying back loans		

farm business. Capital may be money or personal items used in the business, or it might be borrowed money. There are two sources of financing: from the owner(s) and from outside parties. Some outside parties are lenders, such as a bank or agricultural credit institution. When the owner invests savings in the farm operation or contributes personal items, such as vehicles, these also are financing activities. Recall that for accounting purposes, the farm business is a separate entity from the owners, so any personal thing used in the farm business is considered a contribution by the owners to the business. **Investing activities** involve buying and selling items needed for production of farm products. Buying and selling items such as land, buildings, machinery, equipment, and breeding livestock are considered investing activities. Agricultural producers are primarily engaged in **operating activities**. These are the day-to-day activities of producing or selling crops and livestock, managing hired help, and so on. Table 1-1 outlines these types of activities.

> **PRACTICE WHAT YOU HAVE LEARNED**   *At this point, you should be able to complete Problem 1-2 at the end of the chapter.*

The financial statements provide reports of each of these activities. The four main financial statements are the Income Statement, the Statement of Owner Equity, the Balance Sheet, and the Statement of Cash Flows. The **Balance Sheet** reports on most of the results of financing and investing activities. The equity section of the Balance Sheet primarily summarizes the financing activities involving the owner. **Equity** refers to the amount of money owed to the owners of the farm business. Another section on the Balance Sheet, called the liabilities section, summarizes the financing activities involving outside parties. **Liabilities** are the amounts of money owed to outside parties (for example, lenders). The **Statement of Owner Equity** reports the financing activities involving the owner in more detail than the Balance Sheet, and it shows the profit or loss of the farm business. The asset section on the Balance Sheet shows the results of the farm's investing activities. The items purchased by the farm business and expected to earn money for the farm are called **assets**. Thus, the Balance Sheet provides a summary of the financing and investing activities by reporting on the assets, liabilities, and equity of the farm business. The financial position of the farm business can be evaluated from the Balance Sheet.

Learning Objective 6 ■ To name the four primary Farm Financial Statements, to describe the general purpose or function of each of the financial statements, and to name the elements contained in each.

TABLE 1-2 ■ *The four primary Farm Financial Statements and their elements.*

Income Statement	Statement of Owner Equity	Balance Sheet	Statement of Cash Flows
Revenues	Net Income	Assets	Operating activities
Expenses	Investments	Liabilities	Investing activities
Gains	Withdrawals	Equity	Financing activities
Losses			

The **Income Statement** summarizes the results of operating activities: money earned by the farm business, called **revenues**, from the production or sale of farm products, and the costs of operating the farm, called **expenses**. **Net Income** or **Net Profit** is the "bottom line" on the Income Statement and the difference between revenues and expenses. The Income Statement also reports the gains and losses. **Gains** are financial benefits from various activities. For example, when a piece of farm equipment is traded in to purchase a new piece of equipment, depending on the terms of the purchase, a gain may result. (To learn how to tell if a gain does occur, see this example in further detail in Chapter 8.) **Losses** are the opposite of gains. The primary purpose of the Income Statement is to help evaluate the financial performance of the farm operation (whether or not a profit was made). Primarily, it reports on the results of operating activities; however, some of the gains and losses reported on the Income Statement also may arise from financing or investing activities.

The fourth financial statement is the **Statement of Cash Flows**. The Statement of Cash Flows reports on all of the activities (financing, investing, and operating) involving cash. The reader of the Statement of Cash Flows can learn where money came from and where it went during the year, for example, how much money was borrowed this year and what it was used for. The Statement of Cash Flows provides details about the farm's money activities for a year.

Table 1-2 depicts the elements of the four primary Farm Financial Statements.

PRACTICE WHAT YOU HAVE LEARNED   *At this point, you should be able to complete Problem 1-3 at the end of the chapter.*

INCOME STATEMENT

Learning Objective 7 ■ To define and calculate "Net Income" or "Net Loss."

The Income Statement summarizes the financial activities from the sale of farm products (revenues), the costs to produce and sell those products (expenses), and gains and losses. When revenues and gains exceed expenses and losses, the difference is reported as Net Income, profit, or earnings. When the reverse is true, the difference is reported as net loss. This relationship is depicted in the following general format for the Income Statement:

Revenue + Gains − Expenses − Losses = Net Income or (Net Loss)

(The parentheses around "Net Loss" indicate that it is a negative number.)

PRACTICE WHAT YOU HAVE LEARNED   *Can you calculate Net Income? See Problem 1-4.*

The farm manager or owner and outside parties want to evaluate the financial performance of the farm business. The Income Statement is the primary financial statement for this evaluation. They need to know how successful the operating activities have been. However, some of the gains and losses on the Income Statement are the result of financing and investing activities. Therefore, Net Income will include the results of operating activities, and the effects of some financing or investing activities. In that case, the general format is inadequate for evaluation purposes.

For that reason, the Income Statement is usually divided into subsections, to separate the operating activities from the financing and investing activities. The operating activities are presented first, followed by the other activities. The following format for the Income Statement is more useful in presenting the results of the operating activities:

Farm Revenues
  − Operating Expenses
  = Operating Income
  + Other Revenues and Gains
  − Other Expenses and Losses
  = Net Income

Each of these items on the Income Statement represents categories of financial items. Each category summarizes the various types of items within each category.

In this format, **Farm Revenues** include the money earned from the production or sale of farm products and gains from operating activities. Examples of Farm Revenues are the money received for the sale of a wheat crop and the cash received from an insurance claim for crop damage. An example of an operating gain is the value of the feeder pigs raised on the farm. **Operating Expenses** is the amount of money paid for supplies and other costs that are required to operate the farm business and any losses from operating activities. Examples of Operating Expenses are feed, fertilizer, herbicides and pesticides, wages for hired help, insurance, and feeder livestock, just to name a few. An example of an operating loss would be a loss from the sale of culled breeding livestock. **Operating Income** is calculated as the difference between the Revenues and the Operating Expenses (Revenues minus Operating Expenses). **Other Revenues** is money earned from other types of farm activities besides the production and sale of farm products and operating gains. An example of Other Revenues is interest earned on a checking account. The Gains listed with Other Revenues are the gains from financing and investing activities, such as the gain on a trade-in of a vehicle. **Other Expenses** are costs associated with financing and investing activities, for example, interest paid on a loan, and farm income taxes. Losses listed with Other Expenses are losses associated with farm financing and investing activities. For example, a trade-in of a vehicle could result in a loss instead of a gain.

> **Exercise 1-2** *Can you name which of the items on the Farmers' Income Statement (Appendix A) are revenues, expenses, gains, and losses? Answer: (See Appendix A).*

The format presented above is only one way to present the Income Statement. You might see an Income Statement with more detail. Sometimes the analysis of the Income Statement requires more detail than the format above. The amount of detail required will depend upon what the farm owner or manager, or the lender, want to evaluate. An alternative version for the Income Statement lists all farm

Learning Objective 8 ■ To state the general format for the Income Statement and to define or describe each of the items or categories in it.

Learning Objective 9 ■ To explain the advantage of the expanded format of the Income Statement and to state the main differences between the general format and the expanded format.

TABLE 1-3 ■ *Expanded format for the Income Statement.*

Gross Revenue
− Operating Expenses +/−Adjustments
− Depreciation Expense
− Interest Expense +/− Change in Interest Payable
= Net Farm Income from Operations
+/− Gains/Losses on the Sale of Farm Capital Assets
+/− Gains/Losses Due to Changes in General Base Values of Breeding Livestock
= Accrual Adjusted Net Farm Income
+ Other Revenue
− Other Expenses
= Income before Taxes
− Income Tax Expense (farm business taxes only) +/− Change in Taxes Payable
= Accrual Adjusted Net Income

Source: Adapted from Farm Financial Standards Council. *Financial Guidelines for Agricultural Producers.* (Naperville, IL: 1997).

revenues and certain adjustments first, followed by several subcategories and subtotals. Table 1-3 presents an expanded Income Statement using this format, which you can refer to while reading the following paragraphs.

Farm Revenues include the following items. This list might not include all sources of revenue for all farm operations, but considers those most commonly found.

- Crop cash sales
- Changes in crop inventories
- Cash sales of market livestock and poultry
- Changes in market livestock and poultry inventories
- Livestock products sales (for example, milk, eggs, wool)
- Proceeds from government programs
- Gain or loss from the sale of culled breeding livestock
- The change in value due to changes in quantity of raised breeding livestock
- Crop insurance proceeds
- Changes in accounts receivable

The sum of these items is reported as **Gross Revenue**.

The next item on the Income Statement is Operating Expenses, the amount of money paid for supplies and other costs that are required to operate the farm business plus or minus certain adjustments (explained in Chapter 4). The next item is **Depreciation Expense**, an item related to annually allocating the cost of buying assets. The next item is **Interest Expense**, the amount of money paid for interest on loans from a lender. This item may also include certain adjustments (also explained in Chapter 4). These items are added and subtracted to calculate an item called **Net Farm Income from Operations**, which is another version of Operating Income. Net Farm Income from Operations differs from Operating Income above because it includes some adjustments for Operating Expenses and Interest Expense.

Then the Income Statement separates the Gains and Losses from the Other Revenues and Other Expenses. (These categories of Gains and Losses are defined in more detail in Chapter 2.) The Gains and Losses are added or subtracted, as the case may be, to calculate **Accrual Adjusted Net Farm Income**. Accrual Adjusted

Net Farm Income is another way to present Operating Income with the addition of these specific Gains and Losses. The Other Revenue and Other Expenses are the same as those defined above, except that Interest Expense is part of Net Farm Income from Operations and is not included here as part of Other Expenses. When Other Revenue and Other Expenses are added to and subtracted, respectively, from Accrual Adjusted Net Farm Income, another subtotal is calculated, which is called **Income before Taxes**. The next item is the amount paid for income taxes, **Income Tax Expense**, plus certain adjustments (explained in Chapter 4). When these items are added or subtracted, **Accrual Adjusted Net Income** is calculated. Accrual Adjusted Net Income is similar to Net Income described above.

The main difference between the expanded format and the general format indicated earlier in this chapter is the level of detail, which is more extensive in the expanded format. The adjustments in the expanded format (see Chapter 4) provide a more accurate assessment of financial performance.

STATEMENT OF OWNER EQUITY

The Statement of Owner Equity reports the changes in the owner's interest in the farm business during a period. The owner's interest in the farm business refers to what the owner has put into the business (in the form of money or other capital) and the amount of profit that the farm business has earned for the owner. The owner's interest in the farm business is also known as Equity. When the farm business earns a profit, the owner's equity increases accordingly. The preparation of financial statements requires that the Income Statement be prepared first so that the Statement of Owner Equity can also report Accrual Adjusted Net Income. Other transactions also affect owner equity.

Learning Objective 10 ◼ To define the concept of "equity."

> PRACTICE WHAT YOU HAVE LEARNED   *You should be able to complete Problem 1-5 at the end of the chapter at this point.*

Learning Objective 11 ◼ To state the general and expanded formats for the Statement of Owner Equity and to define or describe each of the items or categories in it.

The general format for the Statement of Owner Equity is as follows:

Equity at the beginning of the year
+ Net Income
+ Investments
− Withdrawals
= Equity at the end of the year

Equity at the beginning of the year is the amount of equity that the owner had when the year began. It comes from the amount of equity that the owner had in the farm business at the end of the previous year. **Investments** represents any additions of cash or other personal assets that the owner (or others, such as family members) contributed to the farm business since the beginning of the current year. Many owners begin their farm operations by contributing some personal items (such as a vehicle) for use in the farm operation. From time to time, the owner may add more money or other personal items to the farm business. The farm owner should keep separate the farm business items and activities from the owner's personal items and activities. This separation is highly recommended so that the financial performance and financial position of the farm business is not compromised by any personal financial activities. Owners or managers of

TABLE 1-4 ■ *Format for Statement of Owner Equity for the Farm Business.*

Owner Equity, beginning of year
Accrual Adjusted Net Income or Net Loss
− Owner Withdrawals
+ Non-Farm Income
+ Other Capital Contributions/Gifts/Inheritances Received
− Other Capital Contributions/Gifts/Inheritances Distributed
= Changes to Retained Capital
+/− Change in Excess of Market Value over Cost/Tax Basis of Farm Capital Assets
+/− Change in Non-Current portion of Deferred Taxes
= Change in Valuation Equity
+/− Total Change in Retained Capital and Valuation Equity
= Owner Equity, end of year

Source: Adapted from Farm Financial Standards Council. *Financial Guidelines for Agricultural Producers.* (Naperville, IL: 1997).

farms and ranches have a need to know the financial status of the agricultural operation as a "stand-alone" entity, even though the owner also owns the assets of the farm. (For example, when personal expenses are included with farm expenses, the farm business may appear less profitable than it really is.) **Withdrawals** is any cash or other assets used by the owner for personal purposes, for example, when the owner uses farm cash to pay for personal expenses. Equity at the end of the year is computed by adding Equity at the beginning of the year to this year's Net Income, plus any additions of cash or other personal assets that the owner contributed this year to the farm business, minus any cash or other assets used by the owner for personal purposes.

Table 1-4 displays an expanded format for the Statement of Owner Equity, which the farm accountant can use when additional information is needed.

In this format, the withdrawals of money by the owners for personal use are called Owner Withdrawals. The next item on the statement, **Non-Farm Income**, refers to any money from other jobs or businesses that the owner earned and contributed to the farm business during the current year. For example, in many farm families, one or more of the owners may have another job to supplement the farm income and to fulfill personal career goals. From time to time, some of this money might be used in the farm business for various expenses instead of using it for personal expenses and activities. When this occurs, the amount used in the farm business should be reported on the Statement of Owner Equity, but should not be reported in the Income Statement because the Income Statement should report only the farm business income. (See Appendix B at the end of this book for more information about Non-Farm Income and related topics.) **Other Capital Contributions/Gifts/Inheritances** are contributions to the owner's equity by the owner, when a personal item is put to use in the farm business, or a contribution from someone other than the farm owners, such as when parents in a farm family transfer the ownership of land or other assets to children as a part of an inheritance. In the year that such events occur, the value of the items contributed can be listed as shown in Table 1-4. Changes in market values of assets (a concept that will become clearer as you study the rest of this book), called **Change in Excess of Market Value over Cost/Tax Basis of Farm Capital Assets** in Table 1-4, result in increases (or decreases) in equity, and should be reported on the Statement of

Owner Equity.[20] Another item in this format represents a tax effect associated with these changes in market values, called **Change in Non-Current Portion of Deferred Taxes** in Table 1-4, and is also reported. Chapters 4 and 9 present discussions of deferred taxes in more detail. Each of these items affects the amount of the owner's interest in the farm business. The Statement of Owner Equity reports the changes to the owner's interest during a given year. In some years there may not be any Non-Farm Income or gifts and inheritances contributed or distributed, so these items would not be reported in that case.

> **Exercise 1-3** *Can you identify each of the equity items in the Farmers' Statement of Owner Equity? How much are the Investments and how much are the Withdrawals? Answer: (See Appendix A).*

BALANCE SHEET

The Income Statement and the Statement of Owner Equity report the results of activities over a specified period, usually one year. A farm's fiscal year can begin at any month of the year or on January 1, whichever is most feasible for the owner to make decisions. At the end of every year, the farm accountant compiles these statements. The next year, a new Income Statement and Statement of Owner Equity are prepared. The Balance Sheet, on the other hand, reports the financial position of the farm operation on a specified date. It reports the values of all of the assets being used in the farm operation, not just the new assets acquired during the past year. Assets disposed of are not included. The Balance Sheet also reports all of the outstanding liabilities (debts not yet paid off). The performance of the farm operation is accumulated in the Equity section of the Balance Sheet and is reported along with the other items that constitute owner equity.

> **Learning Objective 12** ■ To state the Accounting Equation and the general format for the Balance Sheet and to define or describe the items or categories in the Balance Sheet.

The Balance Sheet contains an equation, called the **Accounting Equation**, which shows a relationship between Assets, Liabilities, and Equity.

Assets = Liabilities + Equity

Equity on the Balance Sheet is the same as the Owner Equity at the end of the year from the Statement of Owner Equity.

The first items listed on the Balance Sheet are the Assets; Current Assets are first, followed by Non-Current Assets. **Current assets** are those assets expected to earn money or produce other benefits within one year. Examples of Current Assets are cash (which is listed first on the Balance Sheet), Receivables (which are cash, goods, or services owed to the farm business), Inventory (feeder livestock, feed, and crops on hand), and Prepaid Expenses (which are expenses paid in advance, such as insurance).

> If the farm business has investments in cooperatives, life insurance, or other types of investments, they are also listed on the Balance Sheet as "Investments" and are typically categorized as non-current assets.

Typical **Non-Current Assets** for a farm business are land, buildings and improvements, machinery and equipment, breeding livestock, perennial crops, and natural resources; they are expected to provide benefits to the operation for more than one year or at some future date.

The next items listed on the Balance Sheet are the Liabilities, usually in the order in which they are expected to be paid off. **Current Liabilities**, listed first, are

---

20. Farm Financial Standards Council. *Financial Guidelines for Agricultural Producers.* (Naperville, IL: 1997).

debts that are expected to be paid off within one year. Examples of Current Liabilities are debts owed to suppliers for purchased items not yet paid for, taxes owed to government agencies, interest, and principal owed to lenders that will be settled within the next year.

Non-Current Liabilities will not be paid in full during the next year but will be paid off over a number of years; these include mortgages and other debts for purchases of assets, also deferred taxes and capital leases. Equity is listed last on the Balance Sheet. It indicates the difference between the Liabilities and Assets, and thus represents what is left over for the owner after all debts are paid off (hence, the owner's interest).

A typical format for a Balance Sheet lists the Assets on the left side of the page and the Liabilities and Equity on the right side. Sometimes Assets are listed at the top of the page, the Liabilities are listed underneath them, and Equity is listed under the Liabilities. The following is a general format for the Balance Sheet using the left side/right side display. Mathematically, the Total Assets equals the Total Liabilities and Equity.

Assets:	Liabilities:
Current Assets:	Current Liabilities:
Cash	Accounts Payable
Receivables	Interest Payable
Inventory	Taxes Payable
Prepaid Expenses	Short-Term Notes Payable
Total Current Assets	Total Current Liabilities
Non-Current Assets:	Non-Current Liabilities:
Land	Real Estate Notes Payable
Buildings	Notes Payable
Machinery and Equipment	Deferred Taxes
Breeding Livestock	Total Non-Current Liabilities
Orchard	Total liabilities
Total Non-Current Assets	Equity
Total Assets	Total Liabilities and Equity

The types of assets and liabilities will vary for each farm operation, so the format displayed here might not apply to every farm or ranch. This format offers the general idea of what to expect when you read a Balance Sheet.

The Balance Sheet usually displays subtotals for various categories of items. For example, the Current Assets are listed as a category with a subtotal. There is also a subtotal for the Non-Current Assets. The subtotals are used in some aspects of financial analysis, which you will learn about in Chapter 10. Similarly, the Liabilities are displayed in two categories, Current and Non-Current, along with subtotals. These subtotals are especially useful because they can be used to assess the debt position of the operation.

PRACTICE WHAT YOU HAVE LEARNED   *At this point, you should be able to complete Problem 1-6 at the end of this chapter.*

Learning Objective 13 ■ To identify the components of equity.

The equity section of the Balance Sheet could contain two categories (not shown above), one for an item, called Valuation Equity, and another for Retained Capital. **Valuation Equity** refers to the differences between cost and market values

FIGURE 1-2 ■ *Components of Farm Equity.*

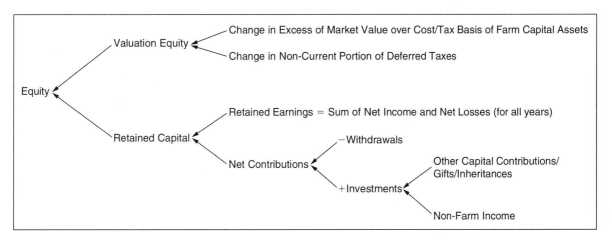

for Non-Current Assets. These would be primarily breeding livestock, land, buildings, machinery and equipment, orchards, and natural resources. **Retained Capital** is the equity from the accumulated Net Income of the farm operation (retained earnings) and the net contributions as seen on the Statement of Owner Equity. Figure 1-2 depicts these items.

The reader of the financial statements can refer to the Statement of Owner Equity to see the details concerning the changes in the owner's interest since the beginning of the current year. The Balance Sheet displays the total Equity at the end of the year, without the details displayed in the Statement of Owner Equity.

Additional details about assets and liabilities are also at the discretion of the farm manager or owner or accountant when the financial statements are prepared. Table 1-5 depicts an example of an expanded, more detailed Balance Sheet for a farm business with several items listed for inventory.

> **Exercise 1-4**   *Can you identify each of the assets, liabilities, and equity items on the Farmers' Balance Sheet? What is the amount of their total assets? What is the amount of their Total liabilities and Total equity? Answer: (See Appendix A).*

STATEMENT OF CASH FLOWS

The Statement of Cash Flows displays the sources and uses of cash for the year and is organized according to the three activities indicated earlier in this chapter. Generally, operating activities involving cash are listed first, followed by cash investing activities, and finally cash financing activities. Any cash transactions that involve the ordinary operations of the business in producing and selling farm products are considered operating activities. Examples of these would be selling crops or livestock and receiving payment for the sale, receiving government payments, paying hired help, paying for seed and fertilizer, paying veterinary bills, and so on. Investing activities would include cash payments for new equipment, land improvements, major building repairs, new breeding livestock, and the sale

Learning Objective 14 ■ To name the three categories of activities reported on the Statement of Cash Flows, to provide examples of each category, and to state the general format for the Statement of Cash Flows.

TABLE 1-5 ■ *Balance Sheet format for a farm business.*

Assets	Liabilities
Cash	Accounts Payable
+ Accounts Receivable	+ Taxes Payable
+ Inventory Raised for Sale	+ Interest Payable
+ Inventory Raised for Use	+ Notes Payable due within one year
+ Inventory Purchased for Resale	+ Real Estate debt due within one year
+ Inventory Purchased for Use	+ Current Deferred Taxes
+ Prepaid Expenses	= Total Current Liabilities
+ Cash Investment in Growing Crops	
= Total Current Assets	
Breeding Livestock	Notes Payable–Non-Current
+ Machinery and Equipment	+ Real Estate Note Payable–Non-Current
+ Buildings and Improvements	+ Non-Current Deferred Taxes
+ Perennial Crops, Orchards, and Natural Resources	= Total Non-Current Liabilities
+ Land	Total Current and Non-Current Liabilities
+ Investments in Cooperatives and Other Investments	
= Total Non-Current Assets	Equity:
	Valuation Equity
	+ Retained Capital
	= Total Equity
Total Current and Non-Current Assets	Total Liabilities and Equity

Source: Adapted from Farm Financial Standards Council. *Financial Guidelines for Agricultural Producers.* (Naperville, IL: 1997).

or trade-in of these assets. The sale of culled breeding livestock is considered an operating activity. Financing activities involve borrowing money and making loan payments to lenders and cash contributions from the owner to the farm business (such as Non-Farm Income), and cash withdrawals from the farm business to the owner for personal expenses. The interest payments, however, are considered an operating activity. The net increase or decrease in cash from all of these activities is added to the amount of cash on hand at the beginning of the year to compute the cash on hand at the end of the year. This amount is the same amount for Cash that is found on the Balance Sheet.

The general format for the Statement of Cash Flows is as follows:

Cash flows from operating activities
+/− Cash flows from investing activities
+ /− Cash flows from financing activities
= Net increase or decrease in cash
+ Cash at the beginning of year
= Cash at the end of the year

The Statement of Cash Flows summarizes information about cash transactions for each of the type of activities involved. The expanded format provides more details about each of the activities. The Income Statement provides a more accurate assessment of the financial performance of the farm business for the year, but the Statement of Cash Flows tells us how cash was used during the year. Table 1-6 displays an expanded format for the Statement of Cash Flows for a farm business.

TABLE 1-6 ■ *Format for the Statement of Cash Flows for the farm business.*

Cash received from sale of livestock

  + Cash received from the sale of crops
  − Cash paid for feeder livestock, purchased feed, and other items for resale
  − Cash paid for all other Operating Expenses
  − Cash paid for interest
  − Net cash paid for taxes
  + Cash received from other miscellaneous farm income
    = Net Cash Provided (or Used) by Operating Activities

Cash received from the sale of breeding livestock (other than culled breeding livestock)

  + Cash received from the sale of machinery and equipment
  + Cash received from the sale of land and buildings
  + Cash received from the sale of investments
  − Cash paid for the purchase of breeding livestock
  − Cash paid for the purchase of machinery and equipment
  − Cash paid for the purchase of land and buildings and improvements
  − Cash paid for the purchase of investments
    = Net Cash Provided (or Used) by Investing Activities

Proceeds from operating loans

  + Proceeds from real estate and other term loans
  + Cash received from gifts and inheritances
  + Cash received from contributions by owners
  − Principal payments for loans
  − Repayments for operating and CCC loans
  − Cash paid out for gifts and inheritances
  − Owner withdrawals
    = Net Cash Provided (or Used) by Financing Activities

Net Increase in Cash from Operating, Investing, and Financing Activities

  + Cash Balance at Beginning of Year
    = Cash Balance at End of Year

Source: Adapted from Farm Financial Standards Council. *Financial Guidelines for Agricultural Producers.*
(Naperville, IL: 1997).

Exercise 1-5 *What is the amount of the cash flows from operating activities on the Farmers' Statement of Cash Flows? How much cash did they use for investing activities? How much cash was provided from their financing activities? Answer: (See Appendix A).*

CHAPTER SUMMARY

Farm Financial Statements provide information about farm financial activities, including financing, investing, and operating activities. The Income Statement reports revenues, expenses, gains, and losses. The Statement of Owner Equity reports in detail the financing activities that pertain to the owner interest in the farm business. It contains information concerning the effects of Net Income, owner withdrawals from and contributions to the farm business, gifts and inheritances received and distributed, and changes in market value of farm assets and the non-current portion of deferred taxes.

The Balance Sheet reports the assets, liabilities, and equity of the farm business. The Balance Sheet lists all Assets, beginning with Current Assets followed by Non-Current Assets; all Liabilities, beginning with Current Liabilities followed by Non-Current Liabilities; and Equity, which may be divided into Valuation Equity and Retained Capital. The Statement of Cash Flows reports on all financing, investing, and operating activities involving cash. The Statement of Cash Flows lists all cash activities classified into operating, investing, and financing categories; additionally, it reconciles the cash balance from the beginning of the year to the end of the year.

BIBLIOGRAPHY

Farm Financial Standards Council. *Financial Guidelines for Agricultural Producers.* Naperville, IL, 1997. Also available online at http://www.ffsc.org.

Financial Accounting Standards Board, "Qualitative Characteristics of Accounting Information", *Statement of Financial Accounting Concepts No. 2* (Stamford, CT: May, 1980).

Financial Accounting Standards Board. "Recognition and Measurement in Financial Statements", *Statement of Financial Accounting Concepts No. 5* (Stamford, CT: 1984).

PROBLEMS

1-1 ■ Matching exercise. Match the following terms in the left column to the definitions in the right column by writing down on a sheet of paper the letter of the definition that most closely describes the terms.

Terms	Definitions
Financial accounting	a. Financial reports that summarize the financial activities of an entity.
Financial performance	b. System that produces reports for outside parties.
Financial statements	c. Procedures to evaluate financial performance and financial position.
Financial position	d. Financial state on a specified date.
Financial analysis	e. Measures of profitability for a given time period.

1-2 ■ Identify each of the following activities as a financing (F), investing (I), or operating (O) activity.

a. Selling hay to a neighbor.

b. Borrowing money for Operating Expenses from Farm Credit Services.

c. Paying the vet bill.

d. Paying back part of the money borrowed from Farm Credit Services.

e. The farm owner decides to use a personal computer for farm record keeping.

f. Spending money from sale of grain to pay off personal credit cards.

g. A young farmer receives 40 acres of land from parents.

h. Buying feed for the entire winter.

i. Buying a new tractor.

j. Buying diesel fuel and storing it on the farm.

k. Selling an old truck.

l. Spouse deposits money from part-time substitute teaching job into farm bank account.

1-3 ▪ On which financial statement would you find the following financial statement categories?

Assets

Equity

Expenses

Gains

Liabilities

Losses

Revenues

1-4 ▪ Calculate Net Income for each of the situations below.

a. Revenues = $1,350,000; Expenses = $950,000.

b. Revenues = $1,350,000; Expenses = $950,000; Gains = $10,000.

c. Revenues = $1,350,000; Expenses = $950,000; Gains = $10,000; Losses = $5,000.

d. Revenues = $1,350,000; Expenses = $950,000; Losses = $35,000.

e. Revenues = $1,350,000; Expenses = $1,450,000; Gains = $10,000.

f. Revenues = $1,350,000; Expenses = $1,450,000; Losses = $10,000.

1-5 ▪ Explain the concept of equity.

1-6 ▪ Identify the category on the Balance Sheet (Current Assets, Non-Current Assets, Current Liabilities, Non-Current Liabilities, Equity) in which each of these items is located.

Accounts Receivable	Fences
Barns	Insurance paid for in advance
Bill owed to utility company	Land
Breeding cattle	Machinery
Cash	Shed
Debt owed to bank for 30-year mortgage on land	Tractors
Feed Inventory	Vehicles

CHAPTER 2

Accounts

Key Terms

Accounts	Current assets	Intangible assets	Posting
Adjunct account	Current liabilities	Journal	Revenue accounts
Automated accounting system	Debit	Journal entry	Single-entry accounting
Balance	Double-entry accounting	Ledger	Source documents
Chart of Accounts	Equity	Manual accounting system	Tangible assets
Contra account	Expense accounts	Non-current assets	Trial balance
Credit	Financial transaction	Non-current liabilities	

You learned in Chapter 1 about an overview of accounting for farm businesses and the activities involved in farm businesses; the nature and purpose of the income statement, statement of owner equity, balance sheet and statement of cash flows; the meaning of terms such as financial position, financial performance, financing activities, investing activities, and operating activities, and how to read farm financial statements.

The next step in understanding and using a farm financial reporting system is to understand the types of items found on the financial statements and how accounting systems manage these items. As mentioned in Chapter 1, the title of each item on the financial statements has a meaning. It is necessary that you understand these definitions. That knowledge and the management of these items adds to the reliability of the statements and helps to compare one farm's financial statements to another's.

In this chapter, you will be able to define the terms "account" and "ledger" and to describe their purpose. You will also learn how to construct the "Chart of Accounts" and to describe its purpose and its construction. You will be able to identify and define the accounts found on farm financial statements. Finally, you will be able to identify the types of source documents used in financial transactions and apply transactions analysis and double-entry accounting in recording transactions.

THE CHART OF ACCOUNTS

The term "account" identifies the means by which farm financial statement items are managed in the financial records. **Accounts** are records of the activities involving all financial statement items. You are probably familiar with having a personal

Learning Objective 1 ■ To define the terms "account" and "ledger" and to describe their purpose.

TABLE 2-1 ■ *Example of cash account for the Farmer's account.*

CASH				
Date	**Description**	**Debits**	**Credits**	**Balance**
Jan. 1	Beginning Balance			20,000
May 12	Purchased repair parts for 750 tractor		250	19,750
July 17	Sold 45 feeder pigs	900		20,650
Oct. 1	Paid principal and interest to bank		11,200	9,450

"bank account." Each person or organization has a separate account in a bank's accounting system. Similarly, each item on farm financial statements has a separate account. Each item's account is a record of the dollar amounts of all of the activities involving that item. The account lists all of the increases and decreases of that item in terms of dollar amounts and provides a net amount or **balance** of all of the increases and decreases. Table 2-1 contains an example of Steve and Chris Farmer's account for Cash.

The top of the account displays the name of the account. The account contains information about activities involving cash. It lists dates and descriptions for all activities in the order in which they occurred. The balance is the amount of cash in the account after each transaction occurs. The amount of cash that belonged to the Farmers' farm business was $20,000 on January 1. On May 12, after the Farmers paid for tractor parts, the amount of cash that belonged to the Farmers' farm business was $19,750, and so on. You will probably notice that this account looks a lot like a checkbook register.

The **ledger** is the collection of all of the accounts for a farm operation. In a **manual accounting system**, in which all accounting is done by hand without the use of a computer and software, each account is listed on a separate ledger page and all the transactions that affect that account are recorded there. In an **automated accounting system**, the same is true, except that it is in a computer and not a book.

Exercise 2-1 *Review the components of an account. What are they? Answer: Name of the account, dates, and descriptions of activities, debits, credits, and balance.*

Learning Objective 2 ■ To define the term "Chart of Accounts" and to describe its purpose and its construction.

The **Chart of Accounts** is the list of all farm financial statement items of significance to a farm operation. Because it is a list of farm financial statement items, it is a list of "accounts." The farm accountant develops the Chart of Accounts when setting up the accounting system for the farm business for the first time, to know what accounts to include in the ledger and what accounts to present on the financial statements. The accountant may modify the chart from time to time as farm activities change.

The Chart of Accounts lists all of the significant farm financial statement items for the farm operation in the order of the five categories discussed in Chapter 1: assets, liabilities, equity, revenues, and expenses. All assets are listed first, then all liabilities, then all equity accounts, and so on. Gains and losses are

TABLE 2-2 ■ *Farm Chart of Accounts.*

1000 Cash
1100 Accounts Receivable
1210 Feeder Livestock Inventory
1220 Feed Inventory
1300 Prepaid Expenses
1600 Machinery and Equipment
1650 Office Furniture and Equipment
1800 Land, Buildings and Improvements
1980 Accumulated Depreciation
2000 Accounts Payable
2100 Taxes Payable
2200 Interest Payable
2300 Notes Payable—Non-Current
2310 Notes Payable Due within One Year
3000 Valuation Equity
3100 Retained Capital
4100 Cash Sales of Market Livestock and Poultry
5000 Feeder Livestock
5020 Purchased Feed
6000 Operating Expenses
7100 Sales Costs
7200 General and Administrative Expenses
8000 Interest Income
8100 Interest Expense
8200 Gains (Losses) on Sales of Farm Capital Assets
8400 Miscellaneous Revenue
9000 Income Tax Expense

also listed, usually after expenses. The Chart of Accounts includes the names of each of the accounts, but may also refer to accounts by number. The numbers assigned to accounts usually follow a logical pattern based on the type of account (asset, liability, equity, revenue, expense, gain, or loss). Appendix C presents the FFSC recommendations for a numbering system for a farm Chart of Accounts and an example based on that system.

The accounts listed in a farm's Chart of Accounts will depend upon the products and activities of the farm operation. It also depends upon the level of detail that the farm manager and farm accountant deem necessary. The Chart of Accounts in Appendix C is very detailed. The farm accountant can adapt the Chart of Accounts in Appendix C to fit the particular operation by adding, deleting, or combining accounts. Table 2-2 shows an example of a Chart of Accounts with less detail. The Chart of Accounts in Table 2-2 would be appropriate for a livestock feeding operation.

To help you learn how to construct a Chart of Accounts, you will work with an example of a 4-H project in this chapter. Some of you might have participated in 4-H projects and may have kept financial records of your project, so you may be able to relate to these procedures.

The preparation of farm financial statements begins by knowing the Chart of Accounts. Steve and Chris Farmer developed a Chart of Accounts when they first set up their accounting system. From the Farmers' financial statements in Appendix A, we can surmise that their Chart of Accounts would look like the following.

Farmers' Chart of Accounts

Cash	Obligations on Leased Assets
Accounts Receivable	Valuation Equity
Feeder Livestock Inventory Raised for Sale	Retained Capital
Feeder Livestock Purchased for Resale	Owner Withdrawals
Feed Inventory Raised for Use	Non-Farm Income
Feed Inventory Purchased for Use	Other Capital Contributions/Gifts/Inheritances
Crop Inventory Raised for Sale	Cash Crop Sales
Prepaid Expenses	Cash Sales of Market Livestock
Cash Investment in Growing Crop	Gains/Losses from Sale of Culled Breeding Livestock
Breeding Livestock	Feeder Livestock
Machinery and Equipment	Purchased Feed
Office Equipment & Furniture	Wage Expense
Perennial Crops	Payroll Tax Expense
Land, Buildings and Improvements	Truck and Machinery Hire
Leased Assets	Herbicides, Pesticides
Accumulated Depreciation	Livestock Supplies, Tools, and Equipment
Accounts Payable	Insurance
Taxes Payable	Real Estate and Personal Property Taxes
Interest Payable	Depreciation Expense
Notes Payable Due within One Year	Interest Expense
Current Deferred Taxes	Gains/Losses on Sales of Farm Capital Assets
Real Estate Notes Payable	Income Tax Expense
Non-Current Deferred Taxes	

DEFINITIONS OF ACCOUNTS

The Chart of Accounts in Appendix C serves as the guide for the definitions and discussion of the accounts below. Many are self-explanatory, but some definitions require discussion.

Assets

Learning Objective 3 ■ To identify accounts for current and non-current assets.

In Chapter 1, you read that assets are the items purchased by the farm business and expected to earn money for the farm. Another way to tell whether an item or account is an asset is to think of assets as those items that the farm business owns or is entitled to. Recall that the term **current assets** designates assets that are either cash, will be sold for cash, or will be used up during the next year. **Non-current assets** are used for more than one year. Non-current assets are sometimes subdivided into two categories: tangible and intangible assets. **Tangible assets** are physical in nature, such as land, buildings, and equipment. **Intangible assets** are nonphysical in nature, such as investments and certain rights, such as a patent. The accounting procedures for tangible and intangible assets are very similar.

Some typical accounts for current assets include the following.

- Cash—record of the amount of cash currently on hand (currency) and in the checking and savings accounts for the farm business.
- Accounts Receivable—record of the amount of money owed to the farm business from the sale of grain, livestock, or other products, which has been finalized but the cash has not been received yet.
- Feeder Livestock Inventory—record of the value of feeder pigs, calves, lambs, poultry, and other meat animals being fed with the intention of selling them. Separate accounts may be set up for feeder livestock that were purchased and feeder livestock that were raised or for each class of meat animal (for example, feeder pigs and feeder calves).
- Breeding Livestock Inventory—record of the value of raised breeding livestock that are for sale, not for use in the herd.
- Feed Inventory—record of the value of the feed currently on hand that will be fed to livestock during the next year. Separate accounts could be set up for purchased feed and grown feed.
- Crop Inventory—record of the value of raised or purchased crops currently on hand that will be sold sometime during the next year.
- Prepaid Expenses—record of costs or expenses paid for in advance and not yet used up. When an expense is "prepaid," another party owes the farm business either goods or services.
- Cash Investment in Growing Crops—record of the annual costs of growing perennial crops (cultivating, pruning, fertilizing, harvesting, and so on) that are in production.[1]

Non-current assets are those assets used in the operation of the farm business that are expected to last more than one year. An account is set up in the Chart of Accounts and in the ledger for each type of non-current asset owned by the farm business. The following are typical accounts for non-current assets.

- Breeding Livestock—record of the value of breeding livestock for use in the herd, not for resale as market animals. Includes beef cattle, dairy cattle, hogs, sheep, poultry, llamas, goats, and horses. Inventory of semen should have a separate account.
- Machinery and Equipment—record of the value of all trucks, tractors, implements, harvesters, and other equipment.
- Office Furniture and Equipment—record of the value of furniture and equipment such as desks, chairs, computers, fax machines, cell phones, and storage cabinets that are used for the farm business.
- Perennial Crops and Natural Resources—record of the costs of planting and developing orchards, groves, vineyards, alfalfa, and bush berries (plants only) during the development period (before production begins).[2] Examples of natural resources would be timberland or gravel pits, revenue from which contributes to farm income. A separate account could be set up for natural resources.

Certificates of deposit are considered investments because they usually result in a penalty for withdrawal before maturity.

If grain is delivered from the farm to the local grain elevator but the check is not received until some time later, the farmer would receive a copy of a document indicating what kind of grain, the total weight or bushels of the grain delivered, the market price on the day of delivery, and the total dollar amount of the sale.

Insurance is a good example of a prepaid expense because it is usually paid at the beginning of the coverage period. The farm business is entitled to coverage by the insurance company during that period, so the farm accountant should record insurance as an asset until it is used up over time.

---

1. Farm Financial Standards Council. *Financial Guidelines for Agricultural Producers.* (Naperville, IL: 1997).
2. Ibid.

- Land, Buildings and Improvements—record of the value of farmland, farm buildings (including sheds and silos), major improvements to buildings, costs of fences, tiling, ditching, water wells, and other improvements to the land.
- Investments in Cooperatives and Other Investments—record of money invested in farm cooperatives and other investments including certificates of deposit, money market funds, mutual funds, stocks, and bonds. Land purchased for the purpose of speculating on land prices, not for farm production purposes, is also an investment.
- Leased Assets—record of the value of assets that are leased from another party and qualify as a capital lease. In a capital lease, the asset is leased for almost all of the asset's normal life or will be owned by the party using the asset at the end of the lease agreement.
- Accumulated Depreciation—an account that shows the total amount of depreciation recorded since the purchase of the assets.

Accumulated Depreciation is not an asset account but is related to non-current assets. This account is an example of a "contra account." A **contra account** is a financial statement item that subtracts from another account. Non-current assets are reported at the original amount that was paid for each asset. Accumulated Depreciation subtracts the amount of the cost that has been allocated over the years that the assets have been owned. (You will learn more about depreciation in Chapters 4 and 6.) A contra account subtracts from another account. An **adjunct account** is a financial statement item that adds to another account. This chapter identifies additional types of contra and adjunct accounts. Chapter 4 provides numerical examples of these accounts.

> **Exercise 2-2** *Now that you can recognize assets, think about a 4-H project, such as a market steer. If you were developing a Chart of Accounts and setting up a ledger, what assets would you have for your market steer project (that is, what asset accounts would you set up in your 4-H accounting system)? Answer: Assets would probably include cash, halters, supplies, clippers, clipping chute, feed, bedding, and market steer. An account would be set up for each of these assets.*

Liabilities

Learning Objective 4 ▦ To identify accounts for current and non-current liabilities.

Liabilities are amounts of money owed to outside parties (for example, lenders) that have not been paid back. From Chapter 1, you know that **current liabilities** are debts that are due within a year. The following are typical accounts for current liabilities in a farm Chart of Accounts.

- Accounts Payable—record of the amount of bills owed to suppliers for purchased goods and services that have not yet been paid for. Examples of unpaid bills would be a bill for the cost of the fuel delivered to the farm or a utility bill received from the power company.
- Taxes Payable—record of the amount of taxes owed but not yet paid to local, state, and federal agencies for payroll, income, property, and other taxes.
- Interest Payable—record of the amount of interest owed to lenders on real estate, equipment, and operating loans.
- Notes Payable Due within One Year—record of the amount of principal on loans (other than real estate) that is due during the next year.

- Real Estate Notes Payable Due within One Year—record of the amount of principal on real estate loans that is due during the next year.
- Current Deferred Taxes—record of taxes that will be due in the future that arise because of differences between financial accounting income and taxable income relating to current assets and current liabilities.
- Obligations on Leased Assets Due within One Year—record of the amount owed to the owner of leased assets that is due within the next year.

Non-current liabilities are debts that will take longer than one year to pay off. In most cases, payments are made every month or every few months or perhaps once every year, until the principal (the loan amount) is paid off. The following are accounts for typical non-current liabilities of a farm operation.

- Notes Payable—record of the amount of principal on loans (other than real estate) that will be paid off after the next year.
- Real Estate Notes Payable—record of the amount of principal owed to lenders on real estate loans that will be paid off after the next year.
- Non-Current Deferred Taxes—record of taxes that will be due in the future that arise because of differences between financial accounting income and taxable income relating to non-current assets. These differences also arise because of differences between market values and costs of non-current assets.[3]
- Obligations on Leased Assets—record of the amount owed to the owner of leased assets for the remainder of the term of the capital lease agreement.

Exercise 2-3 *Think about your 4-H market steer project. Can you identify any liabilities? Think about something you might have purchased for the project but might not be able to pay back until you sell the market steer. Answer: Every 4-H member's situation is different, just as every farm operation is different, but liabilities might include accounts payable. Accounts payable would be a record of the amount of money owed to the feed store for feed, for example, or for the grooming supplies that you bought. Did you borrow money from your parents or a bank to buy the steer? If so, you would also have an account for notes payable for the amount of the loan. You might also have an account for the interest payable (for the amount of interest that you owe on the loan). You would have a balance in these accounts until you paid off the amount that you owe.*

Equity

Equity represents the value of the farm operation to the owner(s).

Learning Objective 5 ■ To identify accounts for equity items.

- Valuation Equity—record of differences between the market value of non-current assets and the original cost of these assets and also the related non-current deferred taxes.[4] These items can be recorded in the following separate accounts:
  - Change in Excess of Market Value over Cost—record of the differences between the original costs of purchased non-current assets and their market values at the end of the year.
  - Change in Non-current Portion of Deferred Taxes—record of the difference in non-current deferred taxes from the end of last year to the end of this year.

---

3. Ibid.
4. Ibid.

■ Retained Capital—record that summarizes all remaining equity transactions. It includes Net Income and the following accounts:

- Owner Withdrawals—record of the amount of farm assets (usually cash) used for personal expenses for the year.
- Non-Farm Income—record of the amount of income from sources other than the farm that was used for farm business transactions rather than for personal expenses.
- Gifts and Inheritances—record of transactions involving transfers of assets to or from the farm operation without payment involved.

> **PRACTICE WHAT YOU HAVE LEARNED** *After studying the definitions for asset, liability, and equity accounts, test yourself by completing Problem 2-1 at the end of this chapter.*

Revenues

Learning Objective 6 ■ To identify accounts for revenues.

Revenue accounts are records of the amount of money earned by the farm operation from the production or sale of farm products. The types of revenue accounts will depend on the types of farm products produced and sold from each farm operation. Some typical accounts for farm revenues include the following.

- Cash Crop Sales—record of the total amount of cash received during the year from selling crops.
- Changes in Crop Inventories—record of the market value of crops and feedstuffs (grain and hay) that were harvested during the year and are still on hand (not yet sold) at the end of the year. This account is also a record of any difference between the purchase price of any crop or feedstuffs that were purchased during the year and the market value of those crops or feedstuffs still on hand at the end of the year. This account is a contra or adjunct account to Cash Crop Sales.
- Cash Sales of Market Livestock and Poultry—record of the total amount of cash received during the year from selling feeder livestock and poultry.
- Changes in Market Livestock and Poultry Inventories—record of the market value of raised market livestock or poultry at the end of the year. This account is also a record of any difference between the purchase price of any market livestock or poultry that were purchased during the year and the market value of that market livestock or poultry still on hand at the end of the year. This account is a contra or adjunct account to Cash Sales of Market Livestock and Poultry.
- Livestock Products Sales—record of the total amount of cash received during the year from selling wool, eggs, milk, and any other livestock products.
- Proceeds from Government Programs—record of the total amount of cash received from government programs during the year that is not intended to be paid back.
- Crop Insurance Proceeds—record of the total amount of cash received during the year from insurance companies for crop damage.
- Gains/Losses from the Sale of Culled Breeding Livestock—record of the difference between the cash amount received from selling culled breeding livestock and the balance sheet value of that livestock on the day that it was sold.

These sales are considered to be the normal annual culling of livestock, not the liquidation of large numbers of livestock that result in downsizing the herd or flock. The farm accountant should record the gains or losses from downsizing or liquidating the herd or flock as Gains/Losses on the Sale of Farm Capital Assets.

- Change in Value Due to Change in Quantity of Raised Breeding Livestock—record of the increase or decrease in the total value of raised breeding stock due to changes in the number of breeding livestock on the farm from the end of last year to the end of this year. This account is also a record of changes in the total value of raised breeding livestock due to increases in age of individual breeding animals from the end of last year to the end of this year.[5]
- Change in Accounts Receivable—record of the difference between the accounts receivable balance at the end of last year and the accounts receivable balance at the end of this year. This account is a contra or adjunct account to all other revenue accounts.

Exercise 2-4   *Think again about your 4-H market steer project. What accounts would you use to record the revenue from the steer? (Hint: Study the definitions very carefully. Your answer depends on whether or not you have sold the steer when you record the revenue). Answer: If you have sold the steer, you would record the amount of cash received in the account entitled Cash Sales of Market Livestock and Poultry. If you have not yet sold the steer, you would record the revenue in the account entitled Changes in Market Livestock and Poultry Inventories. The amount of revenue that you would record depends on whether you raised the steer or whether you bought the steer when you started the project. If you bought the steer, you would record the difference between the purchase price and the market value. If you raised the steer, you would record the market value of the steer.*

Expenses

Expense accounts are records of the costs incurred by the farm for the year. Some expenses involve operating activities and others relate to financing and investing activities. Accounts are set up in the Chart of Accounts and the ledger for each individual expense or group of expenses. The farm accountant decides which accounts to set up depending upon the type of farm operation, the types of costs incurred in the farm business, and the amount of detail that the farm accountant chooses to report. Table 2-3 summarizes the types of expense accounts that can be set up in the Chart of Accounts and the ledger.

The following is an explanation of various types of accounts for production expenses.

- Feeder Livestock—record of the total amount of cash spent during the year for the purchase of feeder calves, feeder pigs, feeder lambs, and poultry. This item includes only the cost of those animals purchased, not raised.
- Purchased Feed—record of the total amount of cash spent for feed during the year.
- Wages Expense—record of the total amount of wages paid for hired help during the year.
- Payroll Tax Expense—record of the total amount of cash paid during the year to the state and federal agencies for the employer's share of Social Security taxes for each of the employees.
- Board for Hired Labor—record of the total amount paid during the year for food, lodging, and any other expenses solely for the benefit of the hired help.

Learning Objective 7 ■ To identify accounts for expenses.

The amount that an employee receives is something less than the total wage because Social Security and income taxes are deducted from the employee's paycheck. However, the farm accountant records the total wage as an expense. The farm business owes the deductions to the state and federal governmental agencies. The farm business must match the amounts that have been deducted from each employee's paycheck for Social Security taxes and pay these amounts also.

---

5. Ibid.

TABLE 2-3 ■ *Types of expense accounts for a farm business.*

Types of expenses	Examples of accounts
Production	Feeder livestock, purchased feed, wages expenses, payroll tax expense, board for hired labor, insurance for hired labor, rent, truck, and machine hire, repairs and maintenance for farm vehicles/machinery/equipment, small tools and supplies, repairs and maintenance for buildings and improvements, fuel, oil, gas, grease, seed, fertilizers, herbicides, pesticides, twine, sacks, poisons, seed tests, veterinarian, vaccinations, medications, breeding fees, registrations, disinfectants, sprays, livestock supplies and tools and equipment, shearing, wool twine and sacks, livestock inspections, office supplies, dues, journals and papers, bank charges, insurance, real estate and property taxes, electricity/water/telephone, depreciation expense, change in accounts payable, change in prepaid insurance, change in cash investment in growing crops.
Selling	Sales costs
General	General and Administrative Expenses
Financing/Investing	Interest Income, Interest Expense, Change in Interest Payable, Gain/Losses on Sales of Farm Capital Assets, Gains/Losses Due to Changes in General Base Values of Breeding Livestock
Other	Miscellaneous Revenue, Income Tax Expense, Change in Taxes Payable, Extraordinary Items

- Insurance for Hired Labor—record of the total amount paid by the farm during the year for any health and accident insurance coverage for the employees.
- Rent—record of the total amount of rent paid during the year to rent land only.
- Truck and Machine Hire—record of the total amount paid during the year for hiring trucks and other machines and equipment, including labor for the operator, if necessary.

Several accounts listed in the Chart of Accounts in Appendix C are self-explanatory, such as Repairs and Maintenance for Farm Vehicles, Machinery, Equipment (parts, labor, tires, and so on), Small Tools and Supplies (hammers, nails, nuts and bolts, welding rods, and so on), and Repairs and Maintenance for Buildings and Improvements (lumber, paint, labor, and so on), as are the accounts for Fuel, Oil, Gas, Grease, Seed, Fertilizers, Herbicides, Pesticides, Twine, Sacks, Poisons, Seed Tests, Veterinarian, Vaccinations, Medications, Breeding Fees, Registrations, Disinfectants, Sprays, Livestock Supplies and Tools and Equipment, Shearing, Wool Twine and Sacks, Livestock Inspections, Office Supplies, Dues, Journal and Papers, and Bank Charges (farm bank account only). Many farm accountants may decide to combine all of these accounts into one account, such as Operating Expenses. Other types of accounts for operating expenses include the following:

- Insurance—record of the total amount paid during the year for all insurance other than health and accident insurance for the employees and personal life insurance.

- Real Estate and Personal Property Taxes—record of the total amount paid during the year for taxes on land and personal property used in the farm business.
- Electricity, Water, and Telephone—record of the total amount paid during the year for utilities that pertain to the farm, not personal use.

Some expense accounts require some calculations and are not necessarily records of cash paid, but are records of some specific production expenses of running the farm operation. Some of these accounts include the following:

- Depreciation Expense—record of the allocation of the cost of purchasing non-current assets. A portion of the original cost of each asset is recorded as depreciation expense each year.
- Change in Accounts Payable—record of the difference between the balance of Accounts Payable at the end of last year and the balance at the end of the current year. This account is a contra or adjunct account of all operating expenses.
- Change in Prepaid Insurance—record of the difference between the balance of Prepaid Insurance at the end of last year and the balance at the end of the current year. This account is a contra or adjunct account of all operating expenses.
- Change in Investment in Growing Crops—record of the difference between the balance of Investment in Growing Crops at the end of last year and the balance at the end of the current year. This account is a contra or adjunct account of all operating expenses.

The preceding list of accounts describes types of operating expenses related to the production of farm products. Other accounts are associated with selling farm products and general expenses. These accounts include:

- Sales Costs—record of expenses related to marketing or selling farm products, such as checkoff fees, transportation, advertising, commissions, storage costs, and financing of storage costs.
- General and Administrative Expenses—record of the cash paid for office labor, office supplies, accounting fees, information systems costs, and so on.

Some accounts report on the expenses associated with financing and investing activities. These accounts include:

- Interest Income—record of the total amount of cash received from interest on investments.
- Interest Expense—record of the cash paid to lenders for the interest portion of loan payments.
- Change in Interest Payable—record of the difference between the balance of Interest Payable at the end of last year and the balance at the end of the current year. This account is a contra or adjunct account to Interest Expense.
- Gain/Losses on Sales of Farm Capital Assets—record of the difference between the cash received for the sale of machinery, equipment, or other non-current asset and the value of that asset on the balance sheet on the day that it was sold.

- Gains/Losses Due to Changes in General Base Values of Breeding Livestock—record of the change in value of breeding livestock due to changes in base values for age categories.[6]

Finally, some revenues and expenses occur that do not readily fall into any specific category. The following accounts are records of these types of items.

Examples of miscellaneous revenue would be tax refunds, sales of timber or other natural resources, and the sale of feed, or breeding livestock (to a neighbor, for example) when the farm operation is not ordinarily in the business of selling feed or breeding livestock to produce revenue. These are unusual or infrequently occurring sources of revenue for the farm.

- Miscellaneous Revenue—record of the cash received from farm income other than those described above.
- Income Tax Expense—record of the total amount of income tax paid on the farm business income only.
- Change in Taxes Payable—record of the difference in the balance of Taxes Payable at the end of last year and the balance at the end of the current year. This account is a contra or adjunct account to Income Tax Expense.
- Extraordinary Items—record of the total amount of income or expense (or gains or losses) considered highly unusual and occurring infrequently.

Exercise 2-5 *Once again, think about the 4-H market steer project. What expense accounts would you set up in your Chart of Accounts and ledger for that project? Think about what costs you would have to pay for to complete the project. Answer: Expense accounts for a market steer project might include Purchased Feed, Vaccinations (if the steer gets sick), Medications (if antibiotics or other feed supplements are added to the feed), Livestock Supplies and Tools (for the cost of the grooming supplies), Depreciation Expense (for depreciation of the clipping chute and clippers), perhaps Sales Costs, Interest Expense (on the loan if you borrowed money to buy the steer), and Gain or Loss on the Sale of Farm Capital Assets (determined when the steer is sold).*

PRACTICE WHAT YOU HAVE LEARNED   *Review the definitions of accounts for revenues, expenses, gains, and losses and complete Problem 2-2 at the end of the chapter.*

RECORDING TRANSACTIONS

Learning Objective 8 To describe the process of transactions analysis.

The management of each account involves recording the increases and decreases of the balances in the accounts as activities (transactions) occur. A **financial transaction** is an activity or event that affects the financial position or financial performance of the company. Every time a financial transaction occurs, it must be analyzed to determine which accounts (from the Chart of Accounts) are affected, and whether there has been an increase or decrease in each account. In every financial transaction, at least two accounts will be affected. For example, when parts for a tractor are purchased for cash, the two accounts that are affected are Repairs and Maintenance Expense and Cash. The Repairs and Maintenance Expense account increases because an additional cost has been incurred in operating the farm, and the Cash account decreases because there is now less cash on hand. Table 2-4 summarizes the steps involved in analyzing financial transactions.

---

6. Ibid.

TABLE 2-4 ■ *Transactions analysis.*

Step 1: Financial transaction occurs.

Step 2: Determine which accounts (at least two from Chart of Accounts) are affected.

Step 3: For each account, determine if the balance increased or decreased.

Transactions Analysis

Step 1: Transaction occurs

Many transactions occur in a farm business each year. You will recognize transactions as being related to the financing, investing, and operating activities described in Chapter 1. (Table 2-5 reproduces the list of activities/transactions from Table 1-1 in Chapter 1.) Every time that farm products are sold, supplies or inventory are purchased, bills are paid, or payments are made on loans or rental agreements, a financial transaction occurs. Each transaction for accounting purposes involves money. Many activities and events occur that do not involve money. An accounting system records only financial transactions (those measured in terms of dollars).

Usually a document provides written evidence of a financial transaction. Accountants often refer to them as **source documents**. A source document is the receipt, invoice, bill, contract, or other paper that accompanies the transaction. Table 2-5 lists some common source documents for each of the types of activities. The source documents provide the details of the transaction for recording. Source documents corroborate the financial statements. If someone wants to check the source of the numbers on the financial statements, these source documents should be organized for easy retrieval if the need arises.

Learning Objective 9 ■ To identify the types of source documents used in financial transactions.

Learning Objective 10 ■ To apply transactions analysis and double-entry accounting in recording transactions.

TABLE 2-5 ■ *Types and examples of farm financial activities and typical source documents.*

Activities	Source documents
Capital contributions	Contracts, titles, checks
Gifts and inheritances	Contracts, titles, checks
—received	
—distributed to heirs	
Non-farm income contributed	Checks written to farm account or for purchases
Withdrawals by owners	Withdrawal slips from bank, bank statements, checks
Borrowing money	Promissory note, check, deposit slip
Paying back loans	Checks, payment schedule from bank
Purchase of assets	Checks, receipts, titles
Sale of assets	Checks, bill of sale
Production/sale of farm products	Receipts, checks, deposit slips
Purchase of inventory	Checks, invoices, bills, statements
Purchase of supplies	Checks, invoices, bills, statements
Purchase of services	Checks, invoices, bills, statements
Payment of wages	Checks, time sheets, employment contracts
Payment of taxes	Checks, bills, statements
Payment of interest	Checks, payment schedule from bank

Step 2: Determine which accounts are affected

Accounting systems are based on the "transactions approach" in which each individual transaction is recorded and summarized in financial statements. The transactions approach contrasts with the "economic approach," which summarizes the effects of the transactions on the financial performance and financial position of a farm business without recording transactions individually. [7]

Every time that a transaction occurs, the source documents are filed and held as evidence for referral when preparing financial statements or tax returns. Each transaction must also be recorded.

To record a transaction, first determine the affected accounts. *Every transaction involves at least two accounts.* For example, suppose that money is borrowed from the bank and deposited in the farm business checking account. The source documents will include the promissory note that indicates how much money (the principal) was borrowed, when it was borrowed, when it has to be paid back, how many payments will be made and the amount of each payment, and the interest rate on the loan. A deposit slip will also be a record that the amount of the loan was deposited in the checking account. The deposit slip is the key document for this transaction.

The preceding transaction is a financing activity. It involves borrowing money, so that means that the farm business now owes money to the bank. When the farm business owes money to an outside party, a liability must be reported on the balance sheet. From the definitions in this chapter, we can identify the accounts that are involved in the transaction. By examining the list of liabilities, we can conclude that the account that most closely describes this liability is Notes Payable. We can be more specific with the account title if we know when it would be paid back (that is, "due within the next year" or non-current). The other account that is involved in the transaction is Cash, because the borrower has deposited money in the farm's checking account.

Exercise 2-6   *The following transactions are typical transactions for the 4-H market steer project. Try to identify the accounts involved in each transaction.*
1. Borrowed $500 to purchase the steer.
2. Used the money to purchase the steer.
3. Bought $100 worth of feed for the steer and charged it.
4. Sold the steer for $1,500.
5. Paid the bill at the feed store.
Answer: 1. Cash and Notes Payable; 2. Cash and Feeder Livestock; 3. (not recorded); 4. Cash and Cash Sales of Market Livestock and Poultry; 5. Cash and Purchased Feed.

Step 3: Determine if the account balance increased or decreased

Each transaction results in a change (an increase or a decrease) in the balance of the affected accounts. In the example above, Cash and Notes Payable are the identified accounts. In this transaction, Cash increased because the farm business received money. Liabilities also increased because the farm business now owes more money than it did before; therefore, Notes Payable also increased.

Exercise 2-7   *For the transactions in Exercise 2-6, some of the account balances increased and some of the account balances decreased. Can you tell which ones increased and which ones decreased? Answer: 1. Both Cash and Notes Payable increased; 2. Cash decreased and Feeder Livestock increased; 3. (none); 4. Both Cash and Cash Sales of Market Livestock and Poultry increased; 5. Cash decreased and Purchased Feed increased.*

---

7. W. Alan Miller and Freddie L. Barnard, "Preparing Reliable Farm Financial Statements: Conceptual and Procedural Issues," *Journal of the American Society of Farm Managers and Rural Appraisers* 60, no. 1 (December 1996): 38–41.

Recording Transactions

Recording a transaction requires that we remember two characteristics. The first is that *every account has two types of events (increases and decreases)*. We can refer to this characteristic as "two sides to every account." The second is that *every transaction has at least two components*. We can refer to this characteristic as "two sides to every transaction." Both characteristics are distinct from each other, but work together to form the procedures for recording transactions.

Two sides to every account

The examples in Step 3 indicate that some transactions result in increases in an account and some transactions result in decreases. Whether the accounting system is manual or computerized, the increases and decreases are shown in two columns—one for the increases and one for the decreases. The left-hand column is known as the **debit** side of the account. When an entry is made on the left side, it is known as debiting the account. The right-hand column is known as the **credit** side of the account. When an entry is made on the right side, it is known as crediting the account.

Two sides to every transaction

In every transaction, the farm accountant records at least one debit and at least one credit. The account that is debited is a different account from the one that is credited. For some accounts, a debit is an increase to the account and for some other accounts it is a decrease to the account. Whether it is an increase or a decrease depends on the type of account it is. The same holds true for credits—sometimes a credit increases an account and sometimes it decreases it, depending on the type of account.

  This way of recording transactions (two sides to every transaction) is known as **double-entry accounting**. Many farm accountants are accustomed to using **single-entry accounting**. Single-entry accounting does not recognize both sides of a transaction. Typically, the increases and decreases in cash are recorded in a check register as each cash transaction occurs. Running balances of other accounts are not maintained, but are updated when financial statements are prepared. In a double-entry system, running balances of accounts are maintained all the time instead of simply being adjusted when financial statements need to be prepared. The double-entry system updates the account balances every time a transaction is recorded so that the farm manager can check the financial position of the farm business at a glance. The double-entry system contributes to the reliability of financial statements because the "two sides to every transaction" maintains a balance for all accounts.

Types of accounts

To determine which account(s) to debit and which account(s) to credit, you need to understand the type of each account involved. The type of account affects whether an increase is on the debit side or the credit side (with the decreases on the opposite side). The following discussion outlines the effect of the type of account on the debits and credits.

- Asset accounts are always increased on the debit side and decreased on the credit side.
- Liabilities and equity accounts are the opposite from assets because they are on the opposite side of the accounting equation; therefore, they are always increased on the credit side and decreased on the debit side.

- Revenues and gains increase the equity in the business, therefore, revenue and gain accounts must also increase on the credit side and decrease on the debit side, just like the equity accounts.
- Expenses, losses, and owner withdrawals decrease equity, so decreases must be on the credit side of these account and increases on the debit side.
- Contra items are recorded on the decrease side of the related account. For example, a contra asset account is recorded on the credit side because it is a decrease of an asset.
- Adjunct accounts are recorded on the increase side of the related account. For example, an item that is adjunct to an expense account is recorded on the debit side because it is an increase of an expense.

Figure 2-1 illustrates these concepts.

In the example of borrowing money, both Cash and Notes Payable increased. Cash is an asset account; therefore, the increase will show up in the debit column, because increases are recorded on the debit side for assets. Notes Payable is a liability account; therefore, the increase will show up in the credit column, because increases are recorded on the credit side for liabilities. That result will not always be the case with every transaction. Sometimes the balance in an account decreases.

FIGURE 2-1 ■ *T-Accounts showing increases and decreases.*

Assets		Liabilities		Equity	
Debits	Credits	Debits	Credits	Debits	Credits
Increases +	Decreases −	Decreases −	Increases +	Decreases −	Increases +

	Revenues/Gains
Debits	Credits
Decreases −	Increases +

	Expenses/Losses/Withdrawals
Debits	Credits
Increases +	Decreases −

Another way to look at it:

Debits	Credits
Increase if account is:	Decrease if account is:
Asset	Asset
Expense	Expense
Loss	Loss
Withdrawal	Withdrawal
Decrease if account is:	Increase if account is:
Liability	Liability
Equity	Equity
Revenue	Revenue
Gain	Gain

Some transactions will result in a decrease in one account and an increase in another account. For example, when buying parts to repair a tractor and paying for it with a check, Cash is decreased (because a payment is made) and Repairs and Maintenance is increased (because an expense for parts occurred). In another case, when a loan payment is made, Cash is decreased and the balance in Notes Payable decreased (because the amount of money owed is less because of the payment). *In every transaction, there is at least one debit and one credit, but increases and decreases depend on what occurs in the transaction and the types of accounts involved.*

Exercise 2-8 *After studying the concepts about debits and credits, use the transactions in Exercises 2-6 and 2-7 and classify the accounts that are involved in each transaction. Then, from Exercise 2-7, knowing whether each account increased or decreased, determine whether each account should be debited or credited.*

Answer:

Account	Type of Account	Increase or Decrease	Debit or Credit
1. Cash	Asset	Increased	Debit
Notes Payable	Liability	Increased	Credit
2. Cash	Asset	Decreased	Credit
Feeder Livestock	Expense	Increased	Debit
3. (None)			
4. Cash	Asset	Increased	Debit
Cash Sales of			
Market Livestock	Revenue	Increased	Credit
and Poultry			
5. Cash	Asset	Decreased	Credit
Purchased Feed	Expense	Increased	Debit

PRACTICE WHAT YOU HAVE LEARNED   *Review the steps involved in the analysis of financial transactions and complete Problem 2-3 at the end of the chapter.*

Journal entries

The **journal** is the name of the book in which transactions are initially recorded. In a manual accounting system, this book is separate from the ledger and in computerized systems, it is a separate section of the software. Transactions are recorded in the journal in the order that they occur. Each recording, called a **journal entry**, consists of the date of the transaction, the accounts involved, the debits and credits, and the amount of money or value involved in the transaction. The transactions analysis determines which accounts are involved, and which accounts to debit and credit. *In every journal entry, the sum of the dollar amounts for the debits must equal the sum of the dollar amounts for the credits.* This equality ensures that the accounting equation remains in balance.

In the journal entry, list the debits first and then the credits, with the credits indented. Returning to the example of purchasing repairs parts for cash, if the cost of parts was $38, the journal entry would look like this:

May 12 Repairs and Maintenance Expense   38
      Cash   38

Notice that the dollar amount for the debit is recorded to the left of the dollar amount for the credit. This format is consistent with how debits and credits are recorded in the accounts. This format identifies the debits and the credits for easy transfer of this information to the accounts in the ledger. If $5,000 was the amount borrowed in the example, it would be recorded in the journal as follows:

May 24 Cash   5,000
        Notes Payable   5,000

After recording the transactions in the journal, the next step is to transfer the information from each journal entry to the ledger. This transfer is called **posting**. In many computerized systems, the posting is done automatically. In a manual system, the farm accountant must perform the posting. For the journal entries prepared above, the information is copied to the relevant accounts in the ledger. After posting, the accounts would contain the information in the following manner (assuming that the Cash account had a balance of $10,995 before the two transactions):

CASH					
Date	**Description**	**Debits**	**Credits**	**Balance**
Beginning Balance				10,995
May 12	Repair parts for 750 tractor		38	10,957
May 24	Borrowed money from bank	5,000		15,957

REPAIRS AND MAINTENANCE EXPENSE				
Date	**Description**	**Debits**	**Credits**	**Balance**
Beginning Balance				0
May 12	Repair parts for 750 tractor	38		38

NOTES PAYABLE				
Date	**Description**	**Debits**	**Credits**	**Balance**
Beginning Balance				0
May 24	Borrowed money from bank		5,000	5,000

Notice that each account has a running balance, so that the farm accountant always knows the balance in each account. The balance is the net difference between the debits and the credits. Normally, asset accounts and expense accounts have more debits than credits (because increases are recorded on the debit side), so these types of accounts should have debit balances. Liability, equity, and revenue accounts normally have more credits than debits (because increases are recorded on the credit side), so these types of accounts should have credit balances.

Trial Balance

The **trial balance** summarizes the accounts in a list of the account titles with their debit or credit balances. A trial balance lists assets first, followed by liabilities, equity accounts, revenues, and expenses, just as in the Chart of Accounts. Assets should have balances in the debit column; liabilities should have balances in the credit column, and so on. A trial balance based on the sample ledger above would appear as follows:

TRIAL BALANCE		
Accounts	Debits	Credits
Cash	15,957	
Notes Payable		5,000
Repairs and Maintenance Expense	38	

The purpose of the trial balance is to display the account balances without having to search through the entire ledger (which can be tedious if many accounts are used in the accounting system), and for checking that the total of all debits is equal to the total of all the credits. If these amounts are not equal, then an error in recording has occurred. The farm accountant would have to examine each entry in the ledger to determine if it was copied from the journal correctly. (The trial balance above is incomplete, so totals are excluded.)

> **Exercise 2-9**   *For the transactions in Exercises 2-6, 2-7, and 2-8, prepare the journal entries and post the entries to ledger accounts. Create an account for Feed Inventory Purchased for Use. This account reports $30 worth of feed left over from last year. Answer: (See Appendix D).*

CHAPTER SUMMARY

The Chart of Accounts lists the farm accounts. We refer to it when determining how to record a transaction. The Chart of Accounts will vary with each farm operation according to the specific activities engaged in by the farm business. Asset accounts are those items owned by the farm business that are producing or will produce future benefits to the farm operation. Liabilities represent obligations to be paid with cash or other assets. Equity represents the value of the farm operation to the owner(s). Revenues are the money earned by the sale of the farm products or the increase in the value of certain assets. Expenses are the costs incurred in the farm business or decreases in the value of certain assets.

Table 2-6 summarizes the activities involved in recording transactions.

TABLE 2-6 ■ *Summary of Accounting Activities.*

1. Financial transactions occur
2. Refer to documents involving the transaction and analyze each transaction according to:
     —What accounts are affected
     —Whether each account has increased or decreased
     —What is the type of each account (asset, liability, equity, revenue, expense)
3. Determine which accounts are debited and credited and record each transaction in the journal
4. Post the journal entries to the ledger
5. Prepare a trial balance and check that debits equal credits and that balances are recorded in the appropriate debit and credit columns

The next chapter provides examples of recording journal entries during the course of a year for a farm operation.

BIBLIOGRAPHY

Farm Financial Standards Council. *Financial Guidelines for Agricultural Producers.* Naperville, IL, 1997. Also available online at http://www.ffsc.org.
Miller, W. Alan, and Freddie L. Barnard. "Preparing Reliable Farm Financial Statements: Conceptual and Procedural Issues." *Journal of the American Society of Farm Managers and Rural Appraisers* 60: 38–41.

PROBLEMS

2-1 ■ Match the following descriptions of asset, liability, and equity accounts with the account names below.

Accounts Payable	a. Record of the value of feedstuffs on hand
Accounts Receivable	b. Record of the value of market livestock on hand
Cash	c. Amount of expenses paid for in advance
Feed Inventory	d. Value of trucks, tractors, implements, harvesters, and so on.
Feeder Livestock Inventory	e. Amount of money used for personal expenses
Machinery and Equipment	f. Amount of money owed to suppliers of goods or services
Notes Payable	g. Amount of money the farm business has
Owner Withdrawals	h. Amount of money owed to lenders
Prepaid Expenses	i. Amount of money owed to the farm business

2-2 ■ Match the following descriptions of revenue, expense, gains, and losses accounts with the account names below.

Cash Crop Sales	a. Difference between balance sheet value of asset sold and the cash received from the sale
Cash Sales of Market Livestock	b. Amount of cash spent for purchase of market livestock
Change in Crop Inventories	c. Amount of cash received from sale of feeder livestock
Change in Market Livestock Inventory	d. Change in value of breeding livestock due to changes in base values
Depreciation Expense	e. Amount of cash received from sale of crops
Feeder Livestock	f. Market value of raised crops still on hand
Gain due to Change in Base Values	g. Amount of cash spent for purchase of feed
Gain due to Change in Quantity	h. Difference between purchase price and current market value of feeder livestock on hand

Income Tax Expense	i. Change in value of breeding livestock due to age progression
Loss on Sale of Farm Capital Assets	j. Amount of income tax paid
Purchased Feed	k. Annual amount of the allocation of purchase price of assets

2-3 ■ For each of the following transactions, complete the table below and identify the accounts involved, the type of each account (asset, liability, equity, revenue, expense), whether each account increases or decreases in the transaction, and whether each account should be debited or credited.

a. Selling hay to a neighbor.

b. Borrowing money for operating expenses from Farm Credit Services.

c. Paying the vet bill.

d. Paying back part of the money borrowed from Farm Credit Services.

e. The farm owner decides to use a personal computer for farm record keeping.

f. Spending money from sale of grain to pay off personal credit cards.

g. A young farmer receives 40 acres of land from parents.

h. Buying enough feed for the entire winter.

i. Buying a new tractor.

j. Buying diesel fuel and storing it on the farm.

k. Selling an old truck.

l. Spouse deposits money from part-time substitute teaching job into farm bank account.

	Accounts	Type of Each Account	Increase or Decrease	Debit or Credit
a.				
b.				
c.				
d.				
e.				
f.				
g.				
h.				
i.				
j.				
k.				
l.				

Journal Entries

In Chapter 2, you learned how to develop a chart of accounts, how to define asset, liability, equity, revenue, expense, gain, and loss accounts, and how to analyze financial transactions.

This chapter continues with the accounting procedures by illustrating typical journal entries for a farm business during a year. You learned in the previous chapter how to record journal entries in the journal and how to post them to the ledger. You were also introduced to the document called the trial balance.

In this chapter, you will learn how to record and post common transactions for financing, investing, and operating activities of farm businesses. You will also learn how to prepare a trial balance. This chapter emphasizes transactions involving cash. The next chapter teaches you about some adjustments that do not involve cash. The maintenance of the cash account is extremely important for predicting cash flows and for preparing the tax return.

FINANCING ACTIVITIES

Some financing activities involve financial transactions with owners of the farm business. Other financing activities involve financial transactions with nonowners such as lenders and other entities. Financing activities result in the farm business owing money to either owners or nonowners.

Learning Objective 1 ■ To record financing activities involving owners.

Investments by the Owners

Investments by owners are the contributions of the business owners to the business. Ordinarily, the owners of a farm business must begin by raising money from outside the business or contributing their own cash or other assets. The farm accountant classifies these transactions as financing activities. Procedures to record investments by the owners include

- Debits to accounts of the assets that are contributed to show the increases.

- Credit to accumulated depreciation for the amount of depreciation on the assets.
- Credit to equity account to indicate the increase in the amount owed to the owners.

Before the year 20X1, Steve and Chris were employed on Steve's family's farm. They owned an old tractor, worth approximately $5,000, and 150 head of breeding cows, worth about $75,000 in market value, both of which were located on Steve's family's farm. A bull that they had purchased previously for $1,000 had an accumulated depreciation of $300. To begin the farm business, they set up a farm bank account and invested their personal savings of $20,000 on January 2, 20X1. They also plan to use their office furniture, worth $1,000, and four-wheel drive pickup truck, worth $15,000 in market value, for the farm business. The result of the transaction analysis will be the following journal entry (1). Notice that account numbers from the Chart of Accounts in Appendix C are listed along with account names for ease in posting.

(1)	Jan. 2	1000 Cash	20,000	
		1650 Office Furniture and Equipment	1,000	
		1600 Machinery and Equipment	20,000	
		1500 Breeding Livestock	76,000	
		1980 Accumulated Depreciation		300
		3100 Retained Capital		116,700

The investment by the Farmers is a transaction involving several accounts. As described in Chapter 2, you must identify, classify, analyze for increases and decreases, and record accounts in a journal entry. Journal entry (1) shows that the owners have transferred the following assets to the farm business: cash, accounts receivable, office furniture, machinery and equipment (a truck and a tractor), and breeding cattle. It also records the amount of the breeding bull that has depreciated (Accumulated Depreciation = $300), which is a credit. The asset accounts are debited for the appropriate dollar amounts and Retained Capital is credited for the net amount of assets and depreciation. Notice that the sum of debits equal the sum of the debits.

If the owners owe money for farm purchases or other liabilities related to the farm business, they must also record these when the farm business is set up.

- Debit is made to equity account to offset the amount owed to owners.
- Credits to liability accounts to show the amounts owed to outsiders.

Suppose that Steve and Chris also have some liabilities at the end of the year 20X0. The liabilities that are relevant for the farm business include taxes owed on farm income from 20X0 in the amount of $2,160 and a debt of $10,000 in the form of a one-year note that existed at the end of 20X0. As part of setting up the farm business, the Farmers record the following entry on January 2:

(2)	Jan. 2	3100 Retained Capital	12,160	
		2100 Taxes Payable		2,160
		2310 Notes Payable Due within One Year		10,000

Non-Farm Income and Owner Withdrawals

Other financing activities by the owners include a) non-farm income used in the farm business, and b) the use (withdrawal) of cash (or other assets) by the owners for personal use. A transaction involving **non-farm income** occurs when a check (or cash) from a non-farm source is deposited in the farm bank account which will be used to pay farm bills or for some type of farm purchase.

- Debit to cash account for non-farm income deposited in farm bank account.
- Credit an equity account for the non-farm income that is deposited.

The farm accountant will record a transaction involving **owner withdrawals** when paying for personal expenses with a check from the farm bank account.

- Debit an equity account for the amount of farm money paid for personal use.
- Credit to cash account for amount paid for personal use.

Suppose Steve and Chris have non-farm income from Chris's part-time job as a staff accountant during tax season for one of the local accounting offices. Each time that Chris transfers money from their personal checking account to the farm checking account (for example, $100 on Feb. 12), she would record a debit to the farm Cash account and a credit to the Non-Farm Income account.

| (3) | Feb.12 | 1000 Cash | 100 | |
| | | 3120 Non-Farm Income | | 100 |

Suppose that Chris transferred $150 from the farm checking account to the personal checking account on March 17. She records the journal entry as follows:

| (4) | Mar.17 | 3110 Owner Withdrawals | 150 | |
| | | 1000 Cash | | 150 |

Gifts and Inheritances

A common transaction in the agricultural industry is the transfer of farm assets to heirs of the owners. (An appraisal of the assets would establish the market values.) This type of transaction is a financial transaction for the farm business because these assets are used in the farm business. If the farm business receives the assets:

- Debits are made to the asset accounts that are given to the farm business to indicate the increases.
- Credit to an equity account to indicate the increase in the amount that the farm business owes to the owners.

If the assets are given to heirs, then:

- Debit to an equity account to indicate the decrease in the amount that the farm business owes to the owners.
- Credits are made to the asset accounts that are given to heirs to indicate the decreases.

Suppose that Steve and Chris inherited the farmstead that they will operate, including 300 acres of land with a market value of $800 per acre and farm buildings and improvements worth $110,000. This event is also a financing activity, and the entry to record this and other similar events involving gifts and inheritances would be as follows:

(5)	Jan. 2	1800 Land, Buildings and Improvements	350,000	
		3130 Other Capital Contributions/		
		Gifts/Inheritances		350,000
		(300 acres $\times$ $800/acre = $240,000)		
		($240,000 for land + 110,000 for buildings = $350,000)		

PRACTICE WHAT YOU HAVE LEARNED *A transaction analysis of financing activities involving owners yields the types of journal entries illustrated by the examples in this section. Review the procedures for transactions analysis in Chapter 2 and the examples in this chapter and complete Problem 3-1 at the end of this chapter.*

Loans

Learning Objective 2 ■ To record financing activities involving nonowners.

Farm operators often borrow money from banks or lending institutions, such as the Farm Credit System, to purchase land or other assets. At times, the farm operation needs a short-term operating loan to pay bills until crops or livestock or other products can be sold. A farm owner who borrows money signs a **promissory note**, a promise to pay back the loan plus interest by a certain date (called the "due date"). The money stays in the farm checking account until the farm owner writes a check when the money is used. The farm accountant refers to the promissory note as the source document when recording the borrowing transaction.

■ Debit the cash account to indicate that the cash account increased when the money was deposited.
■ Credit a liability account to indicate an increase in the amount owed to outsiders.

Suppose that Steve and Chris decide to expand their farm and purchase additional land. The local agricultural lender has agreed to lend them $120,000 in a real estate loan to purchase the additional land. The transaction involving the borrowing of the cash on March 1 is recorded as follows:

(6)	Mar. 1	1000 Cash	120,000	
		2400 Real Estate Notes Payable—Non-current		120,000

Suppose that Steve and Chris decide to buy a tractor (for $50,000), and they have to borrow the money from an agricultural lender. The process of recording the transaction for borrowing the money is similar to the real estate example above, except that the liability account could be different than the one used for real estate. The result is the same—a liability account and the cash account are increased:

(7)	Mar. 1	1000 Cash	50,000	
		2310 Notes Payable Due within One Year		50,000

Loan Payments

When the farm owner pays back the loans, the amount of cash paid includes the principal (the amount borrowed) plus interest on the loan. Lenders charge interest at a rate specified in the promissory note. The amount of interest paid is calculated as the interest rate multiplied times the principal.

- Debit the liability account that was used to record the loan when it was borrowed to show that the liability has decreased.
- Debit an account for interest to show that an expense for interest has occurred.
- Credit the cash account for the total amount of interest and principal that is paid.

Steve and Chris had borrowed $10,000 on October 1, 20X0 at an annual interest rate of 12 percent. They paid back the loan on October 1, 20X1, including $1,200 in interest ($10,000 times 12 percent). The payment is recorded as follows:

(8)	Oct. 1	8100 Interest Expense	1,200	
		2310 Notes Payable Due within One Year	10,000	
		1000 Cash		11,200

Chapters 4 and 6 discuss interest and principal payments in more detail.

> **PRACTICE WHAT YOU HAVE LEARNED** *Borrowing money and paying back loans are necessary activities for most farm operations. Knowing how to keep track of loans and payments is essential. Review the preceding examples and test your knowledge of accounting for financing activities involving nonowners by completing Problem 3-2 at the end of this chapter.*

Learning Objective 3 ■ To record investing activities.

INVESTING ACTIVITIES

Purchases of Assets

Investing activities involve the purchase and sale of capital assets (land, machinery, equipment, vehicles, and land improvements). Farm owners can purchase assets in several different ways—auction sales, one-on-one transactions with the seller, a dealer or a real estate broker, and the contractor who builds a building. The bill of sale or the check written at purchase are the primary source documents for recording the purchase. Most of the transactions require a cash payment, although in some cases the seller is willing to finance the purchase (as with the purchase of a truck from a car dealership). When the purchase is financed by the seller, the farm owner signs a promissory note similar to one that is signed at the bank for a bank loan.

- Debit the accounts for assets that are purchased to record the increases in assets.
- Credit the cash account to indicate the decrease in cash.
- Credit a liability account for the purchase price with the seller.

In journal entries (6) and (7), Steve and Chris borrowed money to acquire additional land and a tractor. The journal entry to record the purchase of land would involve an increase in the account for Land and a decrease in Cash:

(9)	Mar. 1	1800 Land, Buildings and Improvements	120,000	
		1000 Cash		120,000

The entries to record the purchase of machinery and equipment, breeding livestock, development of perennial crops and natural resources, and investments are recorded in a similar way using the appropriate accounts.

Sales and Exchanges of Assets

Investing activities also involve the sale or trade-in of capital assets. When a trade-in transaction occurs, an asset (for example, a truck) is "sold" to a dealer as part of the payment on a new asset. Dealers will often offer a trade-in allowance, which reduces the amount of cash that has to be paid. Usually a gain or loss occurs when an asset is sold or traded. Gains occur when the cash received in a sale or the market value of the new asset in a trade-in exceeds the book value of the sold or traded asset. Losses occur when the opposite is true. (Book value is the difference between the original cost of the asset and the accumulated depreciation.)

- Debit accounts for whatever was received from the sale or trade-in (cash or a new asset) to show the increase in assets.
- Credit the account of the old asset that was sold or traded in to show the decrease in that asset.
- Debit the account for accumulated depreciation for the old asset. (This entry eliminates the depreciation associated with the old asset.)
- Credit the cash account in a trade-in for the amount of the cash paid (the purchase price of the new asset minus the trade-in allowance).
- Debit a loss account if a loss occurs on the sale or trade-in.
- Credit a gain account if a gain occurs on the sale or trade-in.

Suppose that Steve and Chris decided to sell the breeding bull to a neighbor. The bull is not being culled, so this transaction qualifies as a sale of an asset. (Culling is an operating activity and is discussed in the next section.) The original cost of the bull was $1,000 and the accumulated depreciation on the bull on May 10 is $300. The Farmers were able to sell the bull to the neighbor for $900. The journal entry to record this sale would be as follows:

(10)	May 10	1000 Cash	900	
		1980 Accumulated Depreciation	300	
		1500 Breeding Livestock		1000
		8200 Gains (Losses) on Sale of Farm Capital Assets		200

The Farmers record the $200 difference between the cash received and the book value of the bull as a gain in the account titled Gain on Sale of Farm Capital Assets. If they sold the bull for only $600, they would record a loss on the sale in the following way:

May 10	1000 Cash		600	
	8200 Gains (Losses) on Sale of Farm Capital Assets		100	
	1980 Accumulated Depreciation		300	
		1500 Breeding Livestock		1000

Trade-ins involve similar procedures. An asset account is debited for the new asset that is purchased and the cash account is credited for the amount that is paid.

Suppose that Steve and Chris traded in the old tractor when they purchased the new one when they borrowed the $50,000 in journal entry (7). The old tractor had a book value of $5,000 and was fully depreciated (which means there is no accumulated depreciation). The dealer would give them only a $4,000 trade-in allowance. The journal entry would record a loss:

(11)	Mar. 1	1600 Machinery and Equipment	50,000	
		8200 Gains (Losses) on Sale of Capital Assets	1,000	
		1600 Machinery and Equipment		5,000
		1000 Cash		46,000

Machinery and Equipment is credited for the book value of the old tractor and the credit to Cash for $46,000 is the amount of the purchase price of $50,000 less the $4,000 trade-in allowance.

Leasing Assets

Some farm operations will lease assets, such as land or equipment, rather than purchase them. A formal lease agreement specifies the lease arrangements. The asset being leased is specified in the lease agreement with the amount of each lease payment and the time period that the asset will be leased (called the "lease term"). Sometimes the farmer who is leasing the asset will pay cash for the lease payment, and sometimes the farmer will forfeit a share of the crop as part or all of the lease payment.

Certain lease arrangements qualify as capital leases. In **capital leases**, the farmer using the asset (the lessee) will own the asset at the end of the lease term. A lease is a capital lease when the agreement specifies that title to the asset will transfer to the lessee at the end of the lease term, or that the lessee can buy the asset for a very low price at the end of the lease term. These types of leasing arrangements are classified as a financing activity because, in essence, they consist of a purchase with installment payments. A capital lease is also an investing activity because an asset is being purchased. However, unlike the purchases described above, the legal ownership of the asset takes place after the asset has been used.

Accounting for capital lease situations requires that the lessee records the asset as Leased Assets, along with a liability for all future lease payments for that asset. The journal entry records this asset and liability at the fair market value of the asset.

- Debit an asset account to indicate the increase in leased assets.
- Credit a liability to indicate that lease payments are owed to the owner of the asset.

The lease agreement will specify the amount of each lease payment, the length of the leasing period, and the amount of the purchase price at the end of the lease. The lease payments are recorded like payments to a bank on a loan (with principal and interest). That is why a capital lease looks like a financing arrangement. Most lease agreements specify that the lessee make the first payment on the day that the lease agreement began. In that case, the first payment contains no interest, but there is an interest component to every subsequent lease payment. For the payments

- Debit the liability account associated with the lease to show the decrease in the amount owed on the lease.
- Debit an account for interest to show that an expense for interest has occurred (not done for the first payment).
- Credit the cash account to show the decrease in cash for the payment.

Suppose that Steve and Chris lease a harvester in the year 20X1 on a lease contract that allows them to own the harvester at the end of the lease term. If the harvester has a fair market value of $100,000, then the Farmers record the acquisition of the harvester in the following way:

Year 20X1:

| (12) | Aug. 1 | 1910 Leased Assets | 100,000 | |
| | | 2600 Obligations on Leased Assets | | 100,000 |

If the lease term is five years and the interest rate on the lease is 10 percent, then the annual lease payments would be approximately $23,982. The first payment made at the beginning of the lease term is a reduction of the obligation:

Year 20X1:

| (13) | Aug. 1 | 2600 Obligations on Leased Assets | 23,982 | |
| | | 1000 Cash | | 23,982 |

Subsequent payments (beginning the following year) would include a portion for interest, which you would record like this:

Year 20X2:

	Aug. 1	8100 Interest Expense	7,602	
		2600 Obligations on Leased Assets	16,380	
		1000 Cash		23,982

At the end of the lease term, the leased asset in a capital lease becomes part of the property of the farm operation that leased it. When this occurs,

- Debit an asset account related to the leased asset to show that the farm operation owns a new asset.
- Credit the Leased Asset account to show the decrease in leased assets.

At the end of five years, when Steve and Chris get title to the harvester, the harvester is reclassified as part of the Machinery and Equipment account. Suppose that the harvester has been depreciated down to a value of $20,000 at the end of the lease term on August 1, 20X6. The journal entry to reclassify the harvester would be as follows:

Year 20X6:

Aug. 1	1600 Machinery and Equipment	20,000	
	1910 Leased Assets		20,000

Investments in Perennial Crops

Farm operators purchase, trade for, or lease most of the assets used in a farm operation, except for raised breeding livestock. Perennial plants, such as orchards and vineyards, are purchased also, but then must be planted and cared for until they begin to produce fruit during the "development phase" of the perennial crop. During the development phase of vineyards and orchards, the costs of purchasing and maintaining the perennial plants are part of the investment in the perennial crop. All of these costs are recorded in an asset account, such as Perennial Crops.

The purchase invoice provides the price of the plants. The costs of planting and caring for the plants, which will primarily be labor and supplies, have to be obtained from invoices from the purchase of supplies and from keeping track of the number of people and the hours and days that they worked in the orchard or vineyard during the development phase. The farm accountant keeps track of any other costs to determine the complete cost of developing the perennial crop.

- Debit an asset account for the purchase price of the plants and every cost that occurs to plant and maintain the plants during the development phase to indicate the increase in investments.
- Credit the cash account when the plants and other costs are paid for.

Suppose that Steve and Chris had invested $45,000 in an apple orchard in the year 20X1. They would record this cost in the following journal entry:

(14)	May 20	1700 Perennial Crops	45,000	
		1000 Cash		45,000

> PRACTICE WHAT YOU HAVE LEARNED   *These examples of purchasing and selling assets follow the general rule for accounting for assets: debits are made when assets are purchased, or otherwise acquired, and credits are made when the assets are sold or disposed of in some other way. With these guidelines in mind, practice the recording of investing activities by completing Problem 3-3 at the end of the chapter.*

OPERATING ACTIVITIES

Learning Objective 4 ■ To record various operating activities involving revenues.

Accounting for operating activities occurs more frequently than accounting for financing and investing because operating activities occur virtually on a daily basis. The following discussion and examples illustrate some of the operating activities of a farm operation and the procedures required to account for them.

Operating activities involve the production and sale of farm products. The purpose of these activities is to produce profit for the farm business. The following are some common activities involving revenue.

Sales of Farm Products

Operating activities include the sale of crops and livestock. Crops or livestock are purchased or raised and sold at the appropriate time. Many of these products are sold in the marketplace (at an auction or a grain elevator) and the price received on the day of the sale will depend on the market price for the type and quality and weight of the animals or grain on that day. When the animals or crops are sold and the money is received, the farm accountant records the amount of cash received in the following way:

- Debit the cash account for increase in money.
- Credit an appropriate revenue account that is descriptive of the type of farm product that was sold to show the increase in money earned from the sale.

Suppose that Steve and Chris purchase some feeder pigs (see journal entry [15] below). In July, they sell half of their feeder pigs for $900. They record the sale of the pigs as follows:

(16)	July 17	1000 Cash	900	
		4100 Cash Sales of Market Livestock		
		and Poultry		900

If Steve and Chris sell part of their grain crop at harvest time for $12,000, they would record the sale in the following way:

| (17) | Aug. 16 | 1000 Cash | 12,000 | |
| | | 4000 Cash Crop Sales | | 12,000 |

If Steve and Chris sold feeder calves for $50,000 on December 1 the entry to record the sale is as follows:

(18)	Dec. 1	1000 Cash	50,000	
		4100 Cash Sales of Market Livestock		
		and Poultry		50,000

Sales of other farm products, such as milk, eggs, wool, and so on, are recorded in a similar way using a suitable account to show the increase in revenue.

Sales of Culled Breeding Livestock

Operating activities also include the selling of culled breeding livestock. Culling livestock is a normal operating activity, even though it involves selling farm assets.

Just like other sold assets, a gain or a loss on the sale may occur. A gain means that revenue has increased and a loss means that revenue has decreased.

- Debit the cash account to show the increase in cash.
- Credit the asset account for breeding livestock to show the decrease in breeding animals.
- Debit a loss account if a loss occurs on the sale.
- Credit a gain account if a gain occurs on the sale.

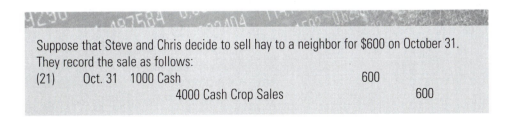

Suppose that Steve and Chris culled five cows, each with a balance sheet value of $300, and sold them for $1,250 on November 20. They record the sale as they recorded the sale of the bull to the neighbor, except that the loss is not recorded as part of the sale of farm capital assets, but rather as part of the sale of culled breeding livestock:

(19)	Nov. 20	1000 Cash	1,250	
		4500 Gains (Losses) on Sale of		
		Culled Breeding Livestock	250	
		1500 Breeding Livestock		1,500

Cash Received from Insurance Companies

Farm operations are subject to crop damage due to bad weather. Most farm operations invest in crop insurance to protect against the loss that occurs when a crop is damaged. When damage occurs, a claim is sent to the insurance company. The cash received from the claim is revenue to the farm operation and the following journal entry is recorded to reflect the increase in revenue.

- Debit the cash account for the amount of the cash received from the insurance company.
- Credit a revenue account (such as Crop Insurance Proceeds).

Sale of Raised Feedstuffs

Operating activities also include selling any crops or feedstuffs that were intended to feed livestock but were sold instead. If any raised feedstuffs were sold instead of being used, the sale would be recorded in a similar way as that of raised crops that are sold.

Suppose that Steve and Chris decide to sell hay to a neighbor for $600 on October 31. They record the sale as follows:

| (21) | Oct. 31 | 1000 Cash | 600 | |
| | | 4000 Cash Crop Sales | | 600 |

Operating activities also include the production of farm products. Many production activities result in costs that must be recorded as expenses. The following are some common activities involving expenses.

Learning Objective 5 ■ To record various operating activities involving expenses.

Exercise 3-1   *Review journal entries (1) through (21) in the examples of the Farmers' transactions. A) Indicate which accounts and dollar amounts would be used to compute Gross Revenue. (Hint: You may want to review the discussion on Gross Revenue in Chapter 1.) B) What other revenue accounts are recorded in journal entries (1) through (21) and where would they be found on the Farmers' income statement? Answer:*

A)

(16) July 17	Cash Sales of Market Livestock and Poultry		$ 900
(17) Aug. 16	Cash Crop Sales		12,000
(21) Oct. 31	Cash Crop Sales		600
(19) Nov. 20	Loss on Sale of Culled Breeding Livestock		(250)
(18) Dec. 1	Cash Sales of Market Livestock and Poultry		50,000
	Gross Revenue		$63,250

Gross Revenue is equal to the sum of the cash sales for market livestock (16) and (18), the crop (17), and the feed (21) minus the loss from the sale of the culled cows (19).

B)

(10) May 10	Gain on Sale of Farm Capital Assets		$ 200
(11) Mar. 1	Loss on Sale of Capital Assets		(1,000)
	Net Loss on Sale of Capital Assets		$ (800)

The gain in (10) and the loss in (11) on the sale and trade-in transactions are combined mathematically to show a net loss of $800, which is subtracted from Net Farm Income from Operations to compute Accrual Adjusted Net Farm Income on the income statement.

Purchases of Market Livestock, Poultry, and Crops

For many farm operations, operating activities include buying market livestock (for example, lambs, feeder pigs, and feeder cattle), market poultry, grain, or other farm products that are sold later. Farm operators purchase these items from various sources (an auction, a neighbor, a dealer, or some other private arrangement). The amount of the purchase is the bid price in an auction or the agreed-upon price in a private deal. This amount is recorded as an expense when the animals or crops are purchased.

■ Debit an expense account that describes the type of livestock or crop that is purchased.
■ Credit the cash account to show that they were paid for.

If Steve and Chris purchased $1,500 worth of feeder pigs, they record the purchase as follows:

(15)	May 16 5000 Feeder Livestock	1,500	
	1000 Cash		1,500

Feeder Livestock is an expense account, representing the amount of expense that was incurred in purchasing the pigs.

Some time later, after the animals are fed or the grain is stored, the products are sold and revenue can be recorded. The sale of half of the Farmers' feeder pigs is illustrated in journal entry (16). Many farms and ranches produce, rather than purchase, market livestock, poultry, crops, or feedstuffs on the farm. The farm accountant records the expenses of raising or producing these products as the costs occur.

Purchase of Feedstuffs

The purchase of feedstuffs (grain, hay, pellets, and so on) is recorded for the amount on the receipt or canceled check.

- Debit an expense account for feed, such as Purchased Feed, or a general account for operating expenses.
- Credit the cash account to show that it was paid for (that there is a decrease in cash because of the payment).

If Steve and Chris purchase $1,000 worth of feed for their livestock on May 10, they record the purchase as follows:

(20)	May 10	5020 Purchased Feed	1,000	
		1000 Cash		1,000

Cash Purchases of Supplies

Many supplies for use on the farm will be paid for when purchased. The farm accountant can consolidate the expenses into a single account, called Operating Expenses or some other descriptive title, when recording these purchases.

- Debit an expense account for the type of supply or a general account for operating expenses.
- Credit the cash account to show that it was paid for (that there is a decrease in cash because of the payment).

Suppose Steve and Chris purchase ear tags for the breeding cattle for $75 on April 17. The journal entry to record the purchase is as follows:

(22)	Apr. 17	6630 Livestock Supplies, Tools, and Equipment	75	
		1000 Cash		75

Similar entries, using suitable accounts, would be recorded for the purchase of other supplies for farm use such as small tools and supplies, fuel, oil, gas, grease, seed, fertilizer, herbicide, pesticide, grain twine and sacks, poisons, seed tests, vaccinations and medications for livestock, disinfectants, sprays, wool twine and sacks, and office supplies.

Annual Costs for Perennial Crops

When vineyards and orchards reach the productive phase (that is, commercial production), the costs that occur for cultivating, grazing, pruning, spraying, plant depreciation, and so on are recorded in expense accounts.

In the first year that the Farmers' orchard reached commercial production, Steve and Chris paid $500 for pesticides. The following journal entry records the cost:

(23)	June 1	6520 Herbicides, Pesticides	500	
		1000 Cash		500

Payments for Parts and Services

Many transactions in a farm business will involve the purchase of various parts and services and are paid for immediately with a check or cash. Expenses for parts and services are recorded in a similar way as the purchase of supplies, feed, or any other expense.

Steve and Chris hire someone to haul the feeder cattle to market on December 1 for $150. They record the cost in the Truck and Machinery Hire account:

(24)	Dec. 1	6310 Truck and Machinery Hire	150	
		1000 Cash		150

An example of a journal entry for the payment of a one-year insurance policy premium is as follows:

(25)	Aug. 1	6700 Insurance	1,200	
		1000 Cash		1,200

Similar entries are recorded for repairs and maintenance of farm vehicles, machinery, equipment, buildings, and other improvements, for rent of farmland, for the cost of veterinary services, breeding fees and registrations, sheep shearing, livestock inspections, insurance, dues, journals and papers, and bank charges.

Wages for Employees

Payroll is the term used to refer to wages paid for hired labor. Most of the time, wages are paid in cash, although some of the benefits to employees are given as room and board. The accounting procedures involve the recording and payment of payroll taxes. The employees pay certain payroll taxes (Social Security, Medicare, and income tax). The farm accountant withholds these taxes from the employees' checks and submits them to the relevant government agency. The employer must also pay Social Security and Medicare taxes, and state and federal unemployment taxes. The following procedures are appropriate for employee wages and payroll taxes:

- Debit an expense account for the gross amount of wages for each employee.
- Credit a liability account for the amount of taxes deducted from each employee's check that the farm operation must submit to the government agencies.
- Credit the cash account for the actual amount of cash paid to the employee.

Suppose that Steve and Chris hire a farmhand on March 1 for $1,200 per month. Each month that the farmhand is paid, they record the gross amount of wages ($1,200) as Wages Expense and the deductions for payroll taxes as Taxes Payable to submit to the government agencies on the employee's behalf. Suppose that the Farmers lived in a state where there is no state income tax and the federal income tax amount is $90. The Social Security and Medicare tax amount would be a percentage (suppose that it is 7.65 percent) of $1,200 (approximately $92). The net pay or cash amount paid to the employee is the gross pay, $1,200, minus the taxes deducted, $90 + 92 = $182. The Farmers' journal entry for a single month's paycheck is as follows:

(26)	Apr. 1	6100 Wage Expense	1,200	
		2100 Taxes Payable		182
		1000 Cash		1,018

The farm accountant must pay the employer's share of Social Security and Medicare taxes plus the unemployment taxes. At least once a year, or perhaps every quarter (four times a year), these taxes must be paid to the relevant government agencies. When the Social Security and Medicare taxes are paid, the employer also pays off the liability for the employee's taxes.

- Debit the liability account for the taxes deducted from the employee's check to show the decrease in liability when the taxes are paid.
- Debit an expense account for the employer's share of the payroll taxes.
- Credit the cash account to show that the taxes have been paid.

When the unemployment taxes are paid:

- Debit an expense account for the unemployment taxes.
- Credit the cash account to show that the taxes have been paid.

The Farmers' share of payroll tax is equal to the amount of Social Security and Medicare taxes that they withheld from the employee's wages, $92 in the previous example. The taxes are due in April for the first quarter of 20X1. Steve and Chris would record the payment in the following manner:

(27)	Apr. 10	2100 Taxes Payable	182	
		6110 Payroll Tax Expense	92	
		1000 Cash		274

This amount paid to the Internal Revenue Service is the sum of $182 deducted from the farmhand's check and the $92 matching amount that Steve and Chris must pay.

Suppose that the state unemployment tax rate is 2.3 percent and the federal unemployment tax rate is 0.8 percent. The Farmers must also pay $27.60 for state unemployment tax ($1,200 times 2.3 percent) and $9.60 for federal unemployment tax ($1,200 times 0.8 percent). They record the payment for the federal unemployment taxes as:

(28)	Apr. 10	6110 Payroll Tax Expense	9.60	
		1000 Cash		9.60

> The payment for the state unemployment taxes are recorded similarly:
>
> (29)   Apr. 10   6110 Payroll Tax Expense   27.60
>   1000 Cash   27.60

Taxes

In addition to payroll taxes, a farm operation is obligated to pay income, real estate, and property taxes.

- Debit an appropriate account for the taxes to record the expense.
- Credit the cash account to show that the taxes have been paid.

> Suppose that real estate and property taxes amount to $1,300 in the year 20X1 for the Farmers and are due on July 1. The journal entry to record these taxes when paid is as follows:
>
> (30)   July 1   6710 Real Estate and Personal Property Taxes   1,300
>   1000 Cash   1,300
>
> The Farmers record the payment of income taxes owed from last year as follows:
>
> (31)   Feb. 1   9100 Income Tax Expense   2,160
>   1000   Cash   2,160

> **PRACTICE WHAT YOU HAVE LEARNED**   *Operating expenses are many and varied for every farm operation. Practice what you have learned about recording operating transactions by completing Problem 3-4 at the end of the chapter.*

POSTING TO THE LEDGER

Learning Objective 6 ■ To post journal entries to a ledger.

The farm accountant analyzes each transaction, records it in the journal, and then posts the journal entry to the ledger, using the procedures outlined in Chapter 2. The farm accountant must be certain that posting occurs immediately following the recording of the journal entry in the ledger. Failure to do so might result in journal entries being omitted from the ledger, which would result in incorrect balances in the accounts.

Figure 3-1 demonstrates the process of posting from the journal to the ledger, using journal entry (31) as an example of the posting process.

> **Exercise 3-2**   *From the journal entries (1) through (31) in this chapter, can you tell how much cash the Farmers have? First, you would have to post all of the journal entries involving cash and then determine the balance in the cash account. Review the ledger in Appendix E and check that each journal entry has been posted. (Hint: The reference numbers for each journal entry are included in the accounts to make it easier.) Answer: $203.80. Verify that this amount for cash is correct by adding up all of the debits in the cash account and subtracting all of the credits.*

FIGURE 3-1 ▪ *Illustration of posting from the journal to the ledger.*

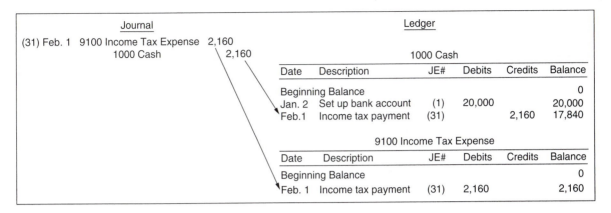

PRACTICE WHAT YOU HAVE LEARNED   *Practice the posting process by completing Problem 3-5 at the end of the chapter.*

At least once a year, a trial balance should be prepared to check for any mistakes in the posting process. The trial balance consists of the debit and credit balances of each account in the ledger. The more frequently the trial balance is prepared, the easier it is to find errors. An error is detected if the totals of debits and credits are not equal to each other. When that happens, the farm accountant must check that each journal entry has been posted correctly. This procedure is easier in a computerized accounting system because the system does most of this without effort.

Before preparing the trial balance, the farm accountant must calculate the account balances (if they are not already calculated). Then, all of the accounts are listed in the order in which they appear in the ledger with one column for the debit balances and one column for the credit balances. To know which balances are debits and belong in the debit column and which are credits and belong in the credit column, the farm accountant must recall which column records the increases for each type of account.

Learning Objective 7 ▪ To prepare a trial balance.

- ▪ Increases are recorded as debits for assets, expenses, and owner withdrawals.
- ▪ Increases are recorded as credits for liabilities, equity, and revenue accounts.

The **normal balance** for each type of account is the debit or credit where increases are recorded.

- ▪ The normal balance for assets, expenses, and owner withdrawals are debits; therefore, the credit entries are subtracted from the debit entries to calculate the balance. If the balance is a positive number, it goes in the debit column on the trial balance. If the result is a negative number, it goes in the credit column on the trial balance.
- ▪ The normal balance for liabilities, equity, and revenue accounts are credits; therefore, the debits are subtracted from the credits to calculate the balance. If the balance is a positive number, it goes in the credit column of the trial balance. If the balance is a negative number, it goes in the debit column of the trial balance.

TABLE 3-1 ■ *Trial balance format.*

Accounts	Debits	Credits
TRIAL BALANCE (Name of Farm Operation) (Date)		
(Assets listed first)	XXXX	
(Liabilities listed next)		XXXX
(Equity accounts listed next)		XXXX
(Revenue accounts listed next)		XXXX
(Expense accounts listed next)	XXXX	
Totals		

Table 3-1 reviews the format for a trial balance, showing that assets are listed first, followed by liabilities, and so on.

The last step is to add all of the numbers in the debit column and then add all of the numbers in the credit column. If the two totals are equal to each other, the trial balance is complete. If they are not equal, the farm accountant must examine all of the journal entries to ensure that they have been recorded and posted properly, and that the balances have been calculated correctly.

Exercise 3-3 *Appendix E displays the Farmers' trial balance that would be prepared after posting journal entries (1) through (31). Check the ledger accounts and make sure that the balance of each account has been correctly placed in the trial balance. (Debit balances should be in the debit column and credit balances should be in the credit column.)*

PRACTICE WHAT YOU HAVE LEARNED *Practice what you have learned about preparing a trial balance by completing Problem 3-6 at the end of the chapter.*

CHAPTER SUMMARY

Journal entries record the increases and decreases in various accounts involved in farm business transactions. This chapter demonstrates how journal entries are recorded during the year in a typical farm operation. Journal entries for financing activities include investments by owners, non-farm income, owner withdrawals, gifts and inheritances, loans, and loan payments. Journal entries for investing activities described in this chapter include the purchase and sale of non-current assets, leasing assets, and investments in perennial crops. Operating activities are numerous for each farm operation. Examples of journal entries included in this chapter are for the purchase and sale of market livestock, poultry, crops, and other farm products; the sale of raised or produced farm products; the sale of culled breeding livestock; the proceeds of crop insurance payments; the purchase of feedstuffs; the sale of raised feedstuffs; the purchase of goods and supplies; the payments for services, wages, and taxes.

PROBLEMS

3-1 ▪ Complete the journal entries for the following financing transactions involving owners.

a. A farmer contributes her personal computer, worth $1,000, to her farm business.

b. A farmer pays a personal credit card debt for $200 with cash earned by the farm business.

c. A farmer deposits $350 from a garage sale of personal items into the farm checking account.

d. A farmer receives $10,000 from a relative's estate sale for use in his farm business.

e. A farmer pays $100 for groceries with the farm checking account.

f. Utilities of $65 for the farmer's house is paid with a check from the farm checking account.

3-2 ▪ Complete the journal entries for the following financing transactions involving nonowners.

a. A farmer borrows $5,000 from the local bank for operating expenses.

b. A farmer makes a payment of $3,500 in interest and $10,000 in principal on a real estate loan.

c. A farmer gets a loan from a relative for $1,000 for use in the farm business.

d. The farmer in (c) pays back the relative without interest.

3-3 ▪ Complete the journal entries for the following investing transactions.

a. A farmer buys a tract of land for $50,000.

b. A farmer leases a tractor that is worth $50,000 and will own the tractor at the end of the lease.

c. A farmer builds a shed for $7,000.

d. A farmer buys five cows for $6,000.

e. A farmer sells a parcel of land to a local municipality for $36,000. The original cost of the land was $36,000.

3-4 ▪ Complete the journal entries for the following operating transactions.

a. A farmer buys feed for $800.

b. A farmer sells feeder calves for $118,000.

c. A farmer buys supplies for the orchard for $36.

d. A farmer pays the $350 veterinary bill.

e. A farmer sells grain for $102,000.

f. A farmer culls his cow herd and sells the cows for $3,000.

3-5 ▪ Post the journal entries from your answers in Problems 3-1 to 3-4. You will have to create a ledger with accounts before you post the entries. Compute the balance in each account.

3-6 ▪ Prepare a trial balance using the ledger accounts that you created in Problem 3-5. Check that the sum of the debits equals the sum of the credits. If they are not equal, check that the journal entries in Problems 3-1 to 3-4 were posted correctly and that the balances in the accounts were computed correctly in Problem 3-5. Finally, make sure that the correct balances have been copied to the correct column in the trial balance.

CHAPTER 4

End of Year Accounting Procedures

Key Terms

Accrual-adjusted approach
Accrual-adjusted financial
  statements
Accrual-basis system
Accrued expenses
Accrued revenues
Adjusted Trial Balance
Adjusting journal entries
Cash-basis system
Change in Accounts Payable

Change in Accounts Receivable
Change in Crop Inventory
Change in Interest Payable
Change in Investment in Growing
  Crops
Change in Market Livestock and
  Poultry Inventory
Change in Prepaid Insurance
Change in Purchased Feed
  Inventory

Change in Purchased Feeder
  Livestock Inventory
Change in Taxes Payable
Change in Taxes Payable
Change in Value Due to Change
  in Quantity of Raised
  Breeding Livestock
Current Deferred Taxes
Deferred tax liabilities
Depreciation expense

Financial accounting systems
Gains/Losses Due to Changes in
  General Base Values of
  Breeding Livestock
Inventory
Non-Current Deferred Taxes
Prepaid expenses
Tax basis
Trial balance
Unadjusted Trial Balance

n Chapter 3, you learned how to record journal entries for transactions involving typical financing, investing, and operating activities in a journal, how to post journal entries to the ledger, and how to prepare an unadjusted trial balance.

The next step in your education of financial accounting is to learn how to prepare farm financial statements. But first, you have to learn about **financial accounting systems**. Financial accounting systems provide the guidelines for preparing financial statements. Under each system, the information reported in the financial statements will vary to some extent.

In this chapter, you are learning about a system that combines two other systems. It is necessary to understand these systems to apply the FFSC Guidelines for farm financial statements. This chapter discusses various adjustments made to farm financial statements at the end of the year. Recall from Chapter 1 that financial statements should be prepared on a periodic and timely basis; therefore, you should perform the adjustments discussed in this chapter at the end of the year or at any other time that you need to prepare financial statements. The adjustments add to the relevance of the financial statements by keeping the statements complete and up-to-date. These adjustments are important if significant in amount, because they may make a difference in decisions. However, the farm accountant has to decide for each of these adjustments if they are significant enough to justify making them. The farm accountant can adjust financial statements directly after they are prepared, or in journal entries posted to the ledger just before preparing the statements. This chapter demonstrates the use of journal entries to record the adjustments. Chapter 5 demonstrates the procedures for adjusting prepared financial statements directly.

In this chapter, you will learn how to describe the differences between the cash-basis system, the accrual-basis system, and the accrual-adjusted approach. This chapter introduces you to two methods for implementing the accrual-adjusted approach. You will learn about one of those methods as you study how to prepare and post adjusting journal entries for inventory, prepaid items, depreciation, accrued expenses, accrued revenues, changes in value of raised breeding livestock, income taxes, and deferred taxes, and how to prepare an adjusted trial balance. Although the accrual-adjusted approach provides up-to-date relevant information, you should realize the importance of maintaining the cash account for predicting cash flows and in preparing the tax return.

ACCRUAL ADJUSTED FARM FINANCIAL STATEMENTS

Learning Objective 1 ■ To define and distinguish between the cash-basis system, the accrual-basis system, and the accrual-adjusted approach.

Many farm operations use the **cash-basis system**, in which the farm accountant records revenue only when cash is received and records expenses only when cash is paid. Cash-basis accounting is relatively easy to apply compared to other systems, so many small businesses and farm operations use it. The journal entries that you learned about and recorded in Chapter 3 are examples of using the cash-basis system.

GAAP requires the use of the accrual-basis system to prepare financial statements. In the **accrual-basis system**, revenue is reported when it is earned, whether the cash has been received or not, and expenses are reported when they occur, not only when they have been paid. The financial statements of an accrual-basis system differ from those of the cash-basis system because sometimes the farm business earns money from the production of crops, livestock, or livestock products before getting paid for them, and because many times the production of crops, livestock, or livestock products costs money before the bills are paid. Figure 4-1 depicts an example of these events.

The accounting difference between the cash-basis and accrual-basis systems is the timing of the reporting of revenues and expenses. Timing becomes an issue if financial statements are prepared between the time of the event and the time that the cash is received or paid. In Figure 4-1, the calendar shows that feeder calves were sold on December 30, but the check was not received until January 2. Similarly, a bill was received on December 26, but was not paid until January. For many farm operations, the financial statements are prepared for each calendar year (on December 31). Under the cash-basis system, the utility bill and the sale of the calves are recorded in January when the check is received and the bill is paid (as you learned in Chapter 3).

FIGURE 4-1 ■ *Calendar 1 events (revenues and expenses).*

Sunday	Monday	Tuesday	Wednesday	Thursday	Friday	Saturday
December 1	2	3	4	5	6	7
8	9	10	11	12	13	14
15	16	17	18	19	20	21
22	23	24	25	26 Utility bill received	27	28
29	30 Feeder calves sold	31	January 1	2 Check received from sale	3 Utility bill paid	4

FIGURE 4-2 ■ *Calendar 2 events (inventory).*

Sunday	Monday	Tuesday	Wednesday	Thursday	Friday	Saturday
December 1	2	3	4	5	6	7
8	9	10	11	12	13	14
15	16	17	18	19	20	21
22	23	24	25	26	27	28
29	30	31	January 1	2	3 Feeder calves sold	4

Under the accrual-basis system, the expense for the utilities and the revenue from the sale of the calves would be recorded in December instead, because that is when the events occurred.

Another difference between the cash-basis and accrual-basis systems is the reporting of inventory. Figure 4-2 depicts a scenario wherein the feeder calves are not sold until January 3. Under the cash-basis system, the expenses for purchasing and raising the feeder calves are recorded during the year when they are paid, but no inventory is reported at the end of the year. The farm operation has nothing to show for the feeder cattle until January, when they are sold and the cash is received. In the accrual-basis system, the balance sheet reports feeder cattle inventory to show that the farm operation did indeed produce something during the year, and the income statement reports the appropriate amount of expenses for the feeder cattle.

The accrual-basis system provides information that is more accurate because the accounting of events occurs in the period in which the events occur. However, the cash-basis system is easier to use because transactions are recorded only when cash changes hands. The accrual-basis system requires much more bookkeeping, because usually more journal entries are recorded. The following journal entries for the utility bill and the sale of feeder cattle from the calendar in Figure 4-1 demonstrate this complexity.

<table>
<tr><td align="center">Cash-Basis</td><td align="center">Accrual-Basis</td></tr>
<tr><td>

When bill is received:
Dec. 26 (no entry)

When bill is paid:
Jan. 3 Utility Expense
      Cash

When calves are sold:
Dec. 30 (no entry)

When check is received:
Jan. 2 Cash
    Sales of Market Livestock

</td><td>

When bill is received:
Dec. 26 Utility Expense
      Accounts Payable

When bill is paid:
Jan. 3 Accounts Payable
      Cash

When calves are sold:
Dec. 30 Accounts Receivable
      Sales of Market Livestock

When check is received:
Jan. 2 Cash
      Accounts Receivable

</td></tr>
</table>

The cash-basis system is easier to apply but provides less accurate information. The accrual-basis system is more complicated, but provides information that is more accurate.

TABLE 4-1 ■ *Journal entries for Calendar 1 events.*

Date and Event	Cash-Basis	Accrual-Basis	Accrual-Adjusted
Dec. 26: Utility bill received	(no entry)	Utility Expense Accounts Payable	(no entry)
Dec. 31: End of year	(no entry)	(no entry)	Change in Accounts Payable Accounts Payable

Jan. 3: When bill is paid	Utility Expense
	   Cash
		Accounts Payable
		   Cash
			Utilities Expense
			   Cash

Dec. 30: When calves are sold	(no entry)	Accounts Receivable Sales of Mkt. Livestock	(no entry)
Dec. 31: End of year	(no entry)	(no entry)	Accounts Receivable Change in Accounts Receivable

Jan. 2: When check is received	Cash
	   Sales of Mkt. Livestock
		Cash
		   Accounts Receivable
			Cash
			   Sales of Mkt. Livestock

To provide a more realistic and accurate picture of the financial position and financial performance of the farm operation than the cash-basis system can provide, the FFSC recommends an alternative to the accrual-basis system, called the **accrual-adjusted approach**. In the accrual-adjusted approach, the farm accountant makes certain adjustments at the end of the year while preparing financial statements. The accrual-adjusted approach minimizes the effort in preparing realistic financial statements without compromising accuracy. The farm accountant makes adjustments only at the end of the year when financial statements are prepared. The difference between the accrual-adjusted approach and the accrual-basis system is that in the accrual-basis system, the types of journal entries that you see above are recorded all year long, but in the accrual-adjusted approach, the adjustments are made only when the financial statements are being prepared. Tables 4-1 and 4-2 display the journal entries for the situations depicted in Figures 4-1 and 4-2.

The accrual-adjusted approach is a hybrid between the cash-basis and accrual-basis systems. The following features characterize the accrual-adjusted approach.

■ End of the year: The accrual-adjusted approach reports events in the year in which the events occur, just as the accrual-basis system does, but the adjustments do not have to be made until the end of the year.

TABLE 4-2 ■ *Journal entries for Calendar 2 events.*

Date and Event	Cash-Basis	Accrual-Basis	Accrual-Adjusted

During the year: Expenses for purchase or raising calves	Expenses
	   Cash
		Mkt. Livestock Inventory
		   Cash
			Expenses
			   Cash

End of year	(no entry)	(no entry)	Mkt. Livestock Inventory Change in Mkt. Livestock Inventory

Jan. 3: When calves are sold for cash	Cash
	   Sales of Mkt. Livestock
		Cash
		   Sales of Mkt. Livestock
		Cost of Sales
		   Mkt. Livestock Inventory
			Cash
			   Sales of Mkt. Livestock

- Only one adjustment: Under the accrual-basis system, several journal entries for Accounts Payable, Accounts Receivable, and other accounts potentially would have to be recorded throughout the year, depending on how often they occur. Only one adjustment is made for these items under the accrual-adjusted approach.
- Inventory: In both the accrual-basis system and the accrual-adjusted approach, inventory is reported in the year in which the inventory is owned, but only one adjustment at the end of the year is required under the accrual-adjusted approach. The cash-basis system does not report inventory.
- In the accrual-adjusted approach, the journal entries for the receipt or payment of cash are identical to those recorded in the cash-basis system, which means that the accrual-adjusted approach uses the cash-basis system, but then it adds the adjustments to approximate the use of the accrual-basis system.

The FFSC recommends the accrual-adjusted approach as an acceptable alternative to the accrual-basis system. This book teaches that approach.

> **PRACTICE WHAT YOU HAVE LEARNED** *To practice your comprehension of the differences among the accrual-basis system, cash-basis system, and the accrual-adjusted approach, complete Problem 4-1 at the end of this chapter.*

In the accrual-adjusted approach, the farm accountant makes adjustments by one of two methods:

Learning Objective 2 ■ To identify two methods for implementing the accrual-adjusted approach.

- Through journal entries recorded before preparing the financial statements
- Directly inserted into the financial statements after they are prepared

When journal entries are used to make the adjustments, they are posted to the ledger (just like the journal entries in Chapter 3). Therefore, the financial statements that are subsequently prepared include the adjustments and are called **accrual-adjusted financial statements**. This chapter presents examples of journal entries to illustrate how the farm accountant records adjustments in the journal. These journal entries are called **adjusting journal entries**.

Under the accrual-adjusted approach, the farm accountant adjusts cash revenues to reflect the amount that the farm operation earned. The farm accountant also adjusts cash expenses to reflect the amount of expenses that occurred. These procedures provide complete, up-to-date, and timely information on the financial statements, and thus contribute to the relevance of the financial statements. The accrual-adjusted approach reports all inventories on the balance sheet and reports inventory adjustments on the income statement. Adjustments are reported as "changes" in relevant accounts and are calculated as the difference in the accounts from the beginning to the end of the year. Figures 4-3 and 4-4 depict the general relationships between cash revenues and cash expenses and related change items.

FIGURE 4-3 ■ *Accrual-Adjusted Revenues.*

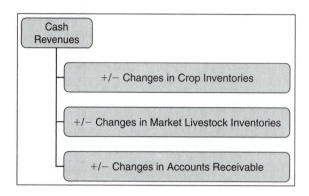

FIGURE 4-4 ■ *Accrual-Adjusted Expenses.*

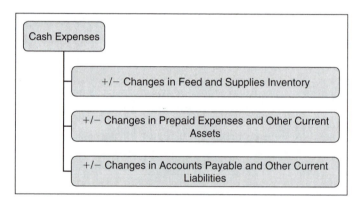

ADJUSTMENTS

This section addresses the following types of adjustments:

- Inventories
- Prepaid expenses
- Depreciation expense
- Accrued expenses
- Accrued revenue
- Income taxes
- Deferred taxes

Inventories

Learning Objective 3 ■ To perform adjustments for various inventories.

Inventory is an asset (purchased or raised) that will be used up or sold within a year as part of the normal operating activities of the farm business. Common types of inventory items for farm operations include feed, crops, and market livestock. Inventory items can be raised or purchased and can be obtained for use on the farm or for resale. Purchased inventory may include the following:

- Purchased feed for use in the operation
- Purchased feed for resale

- Purchased crops for resale
- Purchased market livestock for resale

Raised inventory may include the following:

- Raised feed for use in the operation
- Raised feed for sale
- Raised crops for sale
- Raised market livestock for sale

The farm accountant records the costs of raising or purchasing the inventory as expenses when the cash is paid for them. After the expenses are adjusted, the total amount of the cost for raising or purchasing the inventory pertains only to any inventory that was sold or used up, and it is recorded in the year in which it was sold or used up. The amount of inventory on hand at the end of the year is determined and reported as an asset on the balance sheet. The procedures for inventories are summarized as follows:

During the year:

- Feed, crops, or market livestock are raised or are purchased at various times during the year.
  - Record cash payments for purchase of inventory and other operating expenses
- Feed is used or sold and crops and livestock are sold in various amounts

For accrual-adjusted financial statements at the end of the year:

- Count the amount of purchased and raised feed, crops, and market livestock remaining and determine the dollar value.
  - Make adjustment:
    - Balance sheet shows amount of inventory on hand
    - Income statement shows amount of inventory used up or sold

In the following examples, you will learn about feed and crop inventories first, followed by market livestock inventories.

Exercise 4-1 *Think about your 4-H steer project. Can you think of any inventory that you might purchase for the project as an expense but might not use up entirely by the end of the project (or the end of the year)? Answer: Possible leftover items could be feed and bedding.*

Purchase of feedstuffs for use in the operation

Under the accrual-adjusted approach, the balance sheet reports the amount of purchased feed in a feed inventory account, such as Feed Inventory Purchased for Use. The income statement reports an adjustment for Change in Purchased Feed Inventory, which is added to or subtracted from Purchased Feed (an expense account that shows the amount of cash paid for feed). **Change in Purchased Feed Inventory** is a contra or adjunct account used with Purchased Feed to report the correct amount of purchased feed expense. The procedures must be applied to each type of purchased feed. If a farm operation has, for example, salt and mineral supplements, purchased grain, pellets, hay, hay cubes, and so on, the inventory amount of each one of these items must be determined separately and then added together for the journal entry.

- ■ Count the amount of purchased feed on hand at the end of the year and determine the dollar value.
  - • Make adjustment:
    - • Balance sheet shows amount of inventory on hand
      - • Feed Inventory Purchased for Use
    - • Income statement shows amount of inventory used up or sold
      - • (Purchased Feed)
      - • +/− Change in Purchased Feed Inventory

In Chapter 3, the Farmers purchased $1,000 worth of feed for their livestock and recorded the transaction in the Purchased Feed account on May 10 in journal entry (20). The Farmers determined that the purchased feed on hand at the end of the year cost approximately $230. They had no purchased feed on hand at the beginning of the year. Change in Purchased Feed Inventory is the difference in the amount of feed from the beginning to the end of the year.

Amount of purchased feed leftover	$230
− Feed Inventory Purchased for Use at beginning of the year	0
= Change in Purchased Feed Inventory	$230

If the Farmers used journal entries to make adjustments, they would have recorded the following journal entry and posted it to the ledger. The financial statements would be accrual-adjusted when they were prepared.

(32)   Dec. 31   1224 Feed Inventory Purchased for Use   230
                        5030 Change in Purchased Feed Inventories   230

On the Income Statement, the amount for Change in Purchased Feed Inventory ($230) offsets the amount in the Purchased Feed account ($1,000). The net balance of these two accounts is $770, which is the cost of the feed used up.

Income Statement:
    Operating Expenses:

Purchased Feed	($1,000)
+ Change in Purchased Feed Inventory	230
Net Effect on Net Farm Income from Operations	($770)

Balance Sheet:
    Assets:

Feed Inventory Purchased for Use	$230

Remember that the amount that shows up on the income statement ($770) is only for the amount of feed that was used up. The income statement account, Purchased Feed, reports the total amount paid for purchased feed during the year. If not all of the feed was used up, then not all of the purchased amount should be reported as an expense on the income statement. By making the adjustment for Change in Purchased Feed Inventory, the appropriate amount of expense is reported, which is the goal of the accrual-adjusted approach.

TABLE 4-3 ■ *Cash-Basis vs. Accrual-Adjusted (feed purchased for use).*

Cash-Basis			Accrual-Adjusted		
Purchase			Purchase		
Purchased Feed	1,000		Purchased Feed	1,000	
Cash		1,000	Cash		1,000
Adjustment when financial statements are prepared			Adjustment when financial statements are prepared		
(none)			Feed Inventory Purchased for Use	230	
			Change in Purchased Feed Inventory		230

The journal entries in Table 4-3 demonstrate the difference between cash-basis accounting and the accrual-adjusted approach for purchased feed.

The cash-basis system does not report any feed inventory. The amount of expense for purchased feed in the cash-basis system is overstated ($1,000). The correct amount of purchased feed cost is $770, the amount that was used up. The accrual-adjusted approach provides a more accurate report of the farm's assets by reporting the feed inventory on hand and the correct amount of the expense for Purchased Feed by reporting the net difference between Purchased Feed and Change in Purchased Feed Inventory.

Exercise 4-2 *Think again about the 4-H steer project. You are preparing the financial statements on the project. The feed for the steer consisted of purchased grain and hay. You had some hay but no grain left over after you sold the steer. How would you report the cost of the feed for the steer? Answer: Examine your records to find out how much you spent for the grain and hay. Suppose that you spent $300 for grain and $150 for hay. Next, you have to estimate the amount of feed left over. You have $0 of grain left over and suppose that you have $25 worth of hay left over. You had $30 worth of grain and no hay on hand when you started the project. Your calculations for Change in Purchased Feed Inventory would proceed as follows:*

	Grain	Hay	Total
Amount of purchased feed left over	$ 0	$ 25	$25
– Feed Inventory Purchased for Use at beginning of the year	– 30	0	–30
= Change in Purchased Feed Inventory	$(30)	25	(5)

The journal entry would be recorded with a debit to Change in Purchased Feed Inventory (because the total is negative 5) and a credit to Feed Inventory Purchased for Use:

Change in Purchased Feed Inventory	5	
Feed Inventory Purchased for Use		5

When you post this journal entry to the ledger, the accounts appear as follows:

FEED INVENTORY PURCHASED FOR USE				
Date	**Description**	**Debits**	**Credits**	**Balance**
Beginning Balance				30
Dec. 31 Adjustment			5	25

CHANGE IN PURCHASED FEED INVENTORY				
Date	**Description**	**Debits**	**Credits**	**Balance**
Beginning Balance				0
Dec. 31 Adjustment		5		5

Your income statement reports the total amount of feed purchased ($300 for grain and $150 for hay = $450) adjusted for the change in purchased feed inventory. Your Balance Sheet reports the amount of purchased feed left over.

Income Statement:

Operating Expenses:	
Purchased Feed	($450)
– Change in Purchased Feed Inventory	(5)
Net Effect on Net Farm Income from Operations	($455)

Balance Sheet:

Assets:	
Feed Inventory Purchased for Use	$25

The net cost of feed is the amount used up: the $30 that was left over from last year and the $450 that was purchased except for the $25 that was left over. $30 + 450 – 25 = $455.

Purchased feed for resale

If the farm operation purchased feed for resale instead for use in the farm operation, the adjustments occur in a similar way for any of the feed still on hand when the financial statements are prepared. The only difference from Feed Purchased for Use might be the specific inventory accounts used. Recall from the chart of accounts (Appendix C) that there can be separate accounts for Feed Inventory Purchased for Use (account number 1224) and Feed Inventory Purchased for Resale (account number 1222). The balance sheet could report the amount on hand for these categories of feed inventory separately using these accounts, or could report a single account for purchased feed if it is not important to differentiate between feed purchased for resale and feed purchased for use.

Purchased crop for resale

When the farm business purchases grain or other crops that are intended to be resold, the purchase would be recorded in an account such as Purchased Crops. The farm accountant makes similar adjustments for any of the crop still on hand when the financial statements are prepared, except that the balance sheet reports Crop Inventory Purchased for Resale (account number 1232) and the contra or adjunct account on the income statement is Change in Crop Inventory.

Raised feedstuffs for use in the operation

Under the accrual-adjusted approach, the balance sheet reports the amount of raised feed on hand in an inventory account, such as Feed Inventory Raised for Use. The adjustment on the income statement is Change in Crop Inventory, which is added to other revenue items to compute Gross Revenue. **Change in Crop Inventory** is a contra account that reflects the change in revenue to the farm operation from raising the crop.

- Count the amount of raised feed remaining and determine the dollar value.
  - Make adjustment:
    - Balance sheet shows amount of inventory on hand
      - Feed Inventory Raised for Use
    - Income statement shows amount of revenue from raising inventory
      - Cash Crop Sales (including the amount of raised feed that was sold)
      - +/− Change in Crop Inventory

Steve and Chris complete the harvest of hay for their cow herd on August 1, 20X1, and the value of the hay is $3,500. At the end of the year when they are preparing financial statements, Steve and Chris determine that the value of the hay still on hand on December 31 is $2,300. They had no raised feed on hand at the beginning of the year, so the Change in Crop Inventory is

Amount of raised feed left over at the end of the year	$2,300
− Feed Inventory Raised for Use at beginning of the year	0
= Change in Crop Inventory	$2,300

If the Farmers used journal entries to make adjustments, they would have recorded the following journal entry and posted it to the ledger. The financial statements would be accrual-adjusted when they were prepared.

(33)	Dec. 31	1223 Feed Inventory Raised for Use	2,300	
		4010 Change in Crop Inventory		2,300

The adjustment on the Income Statement reflects the increase in revenue that occurred in 20X1 for the raised feed still on hand. Change in Crop Inventory affects the computation of Gross Revenue as follows:

Income Statement:

Cash Crop Sales	$XXX
+ Change in Crop Inventory	2,300
Effect on Gross Revenue	$2,300

Balance Sheet:
  Assets:

Feed Inventory Raised for Use	$2,300

The income statement reports the total amount of revenue from raising feed. If Change in Crop Inventory is a positive number, it means that the amount of raised feed that is left over at the end of the current year is more than the amount that was left over at the end of last year. That means that some of the feed that was raised was not used up. The income statement account, Cash Crop Sales, reports money received for crops that were sold. If not all of the raised crop was sold, then Cash Crop Sales is not all of the revenue earned by the farm operation. By making the adjustment, the appropriate amount of revenue earned is reported, which is the goal of the accrual-adjusted system.

TABLE 4-4 ■ *Cash-Basis vs. Accrual-Adjusted (feed raised for use).*

Cash-Basis	Accrual-Adjusted	
Harvest	Harvest	
(none)	(none)	
Adjustment when financial statements are prepared	Adjustment when financial statements are prepared	
(none)	Feed Inventory Raised for Use	2,300	
	Change in Crop Inventory	2,300

The journal entries in Table 4-4 demonstrate the difference between the cash-basis system and the accrual-adjusted approach for raised feed for use.

The cash-basis system does not report any feed inventory. The income statement reports only the expenses for producing the feed as part of the operating expenses. Furthermore, the amount of revenue reported in the cash-basis system is understated because the Change in Crop Inventory is not reported. The result is a mismatching of revenues and expenses, because expenses are recorded but no revenues are recorded. The accrual-adjusted approach provides a more accurate report of the farm's assets by reporting the feed inventory on hand and the correct amount of the revenue for raised feed by reporting the Change in Crop Inventory.

Raised feed for sale

If the farm operation raises feed for sale instead for use, the farm accountant makes similar adjustments for any of the feed still on hand when the financial statements are prepared. The only difference from journal entry (33) is the specific inventory account. Recall from the chart of accounts (Appendix C) that there can be separate accounts for Feed Inventory Raised for Sale (account number 1221) and for Feed Inventory Raised for Use (account number 1223). The balance sheet could report the amount on hand for these categories of feed inventory separately using these accounts, or could report a single account for raised feed inventory if the distinction between feed raised for use and feed raised for sale is not necessary.

Raised crops for sale

The farm accountant makes similar adjustments for crops raised for sale. The balance sheet reports any stored crop as Crop Inventory Raised for Sale. On the income statement, Change in Crop Inventory adjusts other revenue items to compute Gross Revenue.

- ■ Count the amount of raised crop on hand and determine the dollar value.
  - • Make adjustment:
    - • Balance sheet shows amount of inventory on hand
      - • Crop Inventory Raised for Sale
    - • Income statement shows amount of revenue from raising inventory
      - • Cash Crop Sales (including the amount of raised crop that was sold)
      - • +/− Change in Crop Inventory

In Chapter 3, Steve and Chris harvested a grain crop and sold part of it for $12,000. They recorded the sale in the Cash Crop Sales account in journal entry (17). The Farmers stored the rest of the crop, valued at $6,000 at the end of the year. If they had no stored crop on hand at the beginning of the year, the Change in Crop Inventory is calculated as follows:

Amount of raised crop left over at the end of the year	$6,000
− Crop Inventory Raised for Sale at beginning of the year	0
= Change in Crop Inventory	$6,000

If the Farmers used journal entries to make adjustments, they would have recorded the following journal entry and posted it to the ledger. The financial statements would be accrual-adjusted when they were prepared.

(34)	Dec. 31	1231 Crop Inventory Raised for Sale	6,000	
		4010 Change in Crop Inventory		6,000

The income statement reflects the additional amount of revenue earned in 20X1 for the harvested crop still on hand. Change in Crop Inventory affects the computation of Gross Revenue as follows:

Income Statement:

Cash Crop Sales	$12,000
+ Change in Crop Inventory	6,000
Effect on Gross Revenue	$18,000

Balance Sheet:
    Assets:

Crop Inventory Raised for Sale	$6,000

      The journal entries in Table 4-5 demonstrate the difference between the cash-basis system and the accrual-adjusted approach for raised crops for sale.

      The cash-basis system does not report any crop inventory. The accrual-adjusted approach provides a more accurate report of the farm's assets by reporting the crop inventory on hand and the correct amount of the revenue for the raised crop by reporting the Change in Crop Inventory.

TABLE 4-5 ▪ *Cash-Basis vs. Accrual-Adjusted (crops raised for sale).*

Cash-Basis			Accrual-Adjusted		
Harvest			Harvest		
(none)			(none)		
Sale			Sale		
Cash	12,000		Cash	12,000	
Cash Crop Sales		12,000	Cash Crop Sales		12,000
Adjustment when financial statements are prepared			Adjustment when financial statements are prepared		
(none)			Crop Inventory Raised for Sale	6,000	
			Change in Crop Inventory		6,000

> PRACTICE WHAT YOU HAVE LEARNED *Practice what you have learned and complete Problem 4-2 at the end of the chapter.*

Purchased market livestock for resale

When the farm operation purchases market livestock for resale, the farm accountant makes adjustments similar to the inventory adjustments for crops and feed. If any of the purchased market livestock is on hand when financial statements are prepared, the balance sheet reports the dollar amount of the purchased livestock in an inventory account, such as Feeder Livestock Inventory Purchased for Resale. The adjustment on the income statement is Change in Purchased Feeder Livestock Inventory. **Change in Purchased Feeder Livestock Inventory** is a contra or adjunct account reported with the Feeder Livestock account to report the correct amount of purchased feeder livestock expense. If the farm operation has more than one type of market livestock (feeder cattle and feeder lambs, for example), the farm accountant performs these procedures separately for each type of animal and then adds the amounts for the journal entry.

- Count the amount of purchased market livestock on hand and determine the dollar value.
  - Make adjustment:
    - Balance sheet shows amount of inventory on hand
      - Feeder Livestock Inventory Purchased for Resale
    - Income statement shows amount of inventory that was sold
      - (Feeder Livestock)
      - +/– Change in Purchased Feeder Livestock Inventory

In Chapter 3, the Farmers purchased feeder pigs for $1,500. They recorded the purchase in journal entry (15) in the Feeder Livestock account on May 16. In July, they sold half of the feeder pigs. The cost of the remaining half of the feeder pigs is $750. At the end of the year, they have not sold the pigs yet, so the Balance Sheet will report Feeder Livestock Inventory Purchased for Resale for $750. The expense of purchasing the feeder pigs is the difference between the $1,500 purchase price and the change in market livestock inventory. If the Farmers had no market livestock on hand at the beginning of the year, the Change in Purchased Feeder Livestock Inventory is calculated as follows:

Amount of purchased market livestock left over at the end of the year	$750
− Feeder Livestock Purchased for Resale at beginning of the year	0
= Change in Purchased Feeder Livestock Inventory	$750

If the Farmers recorded journal entries to make adjustments, they would have recorded the following journal entry and posted it to the ledger. The financial statements would be accrual-adjusted when they were prepared.

(35) Dec 31   1212 Feeder Livestock Inventory Purchased for Resale   750
                  5010 Change in Purchased Feeder
                     Livestock Inventory   750

On the Income Statement, the amount for the Change in Purchased Feeder Livestock Inventory ($750) offsets the amount for Feeder Livestock ($1,500). The net balance of these two accounts is $750, which is the cost of the half of the pigs that were sold.

Income Statement:
    Operating Expenses:

Feeder Livestock	$(1,500)
+ Change in Purchased Feeder Livestock Inventory	750
Net Effect on Net Farm Income from Operations	(750)

Balance Sheet:
    Assets:

Feeder Livestock Inventory Purchased for Resale	$750

The journal entries in Table 4-6 demonstrate the difference between cash-basis accounting and the accrual-adjusted approach for the purchase and sale of market livestock.

The cash-basis system does not report any market livestock inventory. The accrual-adjusted approach provides a more accurate report of the farm's assets by reporting the market livestock inventory on hand and the correct amount of the expense for purchased livestock by reporting the net amount of Feeder Livestock and Change in Purchased Feeder Livestock Inventory.

Raised market livestock for sale

The farm accountant makes similar adjustments for raised market livestock. The balance sheet reports the value of the raised market livestock as Feeder Livestock Inventory Raised for Sale. The adjustment on the income statement is Change in

TABLE 4-6 ■ *Cash-Basis vs. Accrual-Adjusted (livestock purchased for resale).*

Cash-Basis			Accrual-Adjusted		
Purchase			Purchase		
Feeder Livestock	1,500		Feeder Livestock	1,500	
Cash		1,500	Cash		1,500
Sale			Sale		
Cash	900		Cash	900	
Sales of Market Livestock		900	Sales of Market Livestock		900
Adjustment when financial statements are prepared			Adjustment when financial statements are prepared		
(none)			Feeder Livestock Inventory		
			Purchased for Resale	750	
			Change in Purchased Feeder		
			Livestock Inventory		750

Market Livestock and Poultry Inventory. The **Change in Market Livestock and Poultry Inventory** is a contra account representing an increase in revenue for the farm business from raising livestock.

- Count the amount of raised market livestock on hand and determine the dollar value.
  - Make adjustment:
    - Balance sheet shows amount of inventory on hand
      - Feeder Livestock Inventory Raised for Sale
    - Income statement shows amount of revenue from raising inventory
      - Cash Sales of Market Livestock and Poultry (including the amount of raised market livestock and poultry that was sold)
      - +/− Change in Market Livestock and Poultry Inventory

In Chapter 3, Steve and Chris sold raised feeder calves for $50,000. Suppose instead that they still had those feeder calves on hand at the end of the year when they needed to prepare financial statements. If they had no feeder livestock on hand at the beginning of the year, the Change in Market Livestock and Poultry Inventory is calculated as follows:

Amount of raised market livestock on hand at the end of the year	$50,000
− Feeder Livestock Inventory Raised for Sale at beginning of the year	0
= Change in Market Livestock and Poultry Inventory	$50,000

If the Farmers recorded journal entries to make adjustments, they would have recorded the following journal entry and posted it to the ledger. The financial statements would be accrual-adjusted when they were prepared.

(36) Dec. 31   1211 Feeder Livestock Inventory Raised for Sale   50,000
                     4110 Change in Market Livestock
                          and Poultry Inventory   50,000

The Income Statement reflects the amount of the revenue that occurred in 20X1 for the raised market livestock still on hand. Change in Market Livestock and Poultry Inventory affects the computation of Gross Revenue as follows:

Income Statement:

Cash Sales of Market Livestock and Poultry	$XXX
+ Change in Market Livestock and Poultry Inventory	50,000
Effect on Gross Revenue	$50,000

Balance Sheet:
   Assets:

Feeder Livestock Inventory Raised for Sale	$50,000

The journal entries in Table 4-7 demonstrate the difference between cash-basis accounting and the accrual-adjusted approach for raised market livestock.

The cash-basis system does not report any market livestock inventory. The accrual-adjusted approach provides a more accurate report of the farm's assets by reporting the market livestock inventory on hand and the correct amount of revenue for the raised livestock by reporting the Change in Market Livestock and Poultry Inventory.

TABLE 4-7 ■ *Cash-Basis vs. Accrual-Adjusted (raised market livestock).*

Cash-Basis	Accrual-Adjusted	
Adjustment when financial statements are prepared	Adjustment when financial statements are prepared	
(none)	Feeder Livestock Inventory	50,000
	Raised for Sale	
	Change in Market Livestock	50,000
	and Poultry Inventory	

> **PRACTICE WHAT YOU HAVE LEARNED**   *Practice what you have learned and complete Problem 4-3 at the end of the chapter.*

Prepaid Expenses

Prepaid expenses are supplies and other purchases paid for in advance and used up over time. They are recorded as expenses when purchased. When financial statements are prepared, the unused amount for each prepaid item is determined and is reported as an asset on the balance sheet. The income statement reports an adjustment to operating expenses to report the accurate cost of the items purchased.

Learning Objective 4 ■ To perform adjustments for prepaid items.

Prepaid Insurance

Farm operations spend a considerable amount for various types of insurance. Sometimes the insurance policies cover extended periods of time, such as a year or perhaps longer. If the annual insurance premium is paid in full when due, the payment represents insurance coverage paid for in advance. The accrual-adjusted approach requires an adjustment to show the amount of unused insurance coverage on the balance sheet. This amount is reported as Prepaid Expenses and represents the amount of insurance coverage that has not yet been used up and is still owed to the farm business from the insurance company. (The farm accountant may decide to name this account "Prepaid Insurance" to differentiate prepaid insurance from other prepaid expenses and create other prepaid accounts.) The adjustment on the income statement is **Change in Prepaid Insurance**, a contra or adjunct account reported with Insurance Expense to show only the amount of insurance coverage used up for the year.

- ■ Determine the amount of unused insurance at the end of the year.
  - Make adjustment:
    - Balance sheet shows amount of unused insurance
      - Prepaid Expenses
    - Income statement shows amount of insurance that was used up
      - (Insurance)
      - +/– Change in Prepaid Insurance

In Chapter 3, Steve and Chris purchased insurance for $1,200 for a one-year policy on August 1. They recorded the purchase in the Insurance account in journal entry (25). By the end of the year, they have not used up all of the coverage—only the coverage from August 1 to December 31. The $1,200 premium averages out to $100 per month ($1,200 divided by 12 months). By December 31, five months of insurance coverage has expired, so only $500 of insurance expense (5 months times $100) should show up on the Income Statement. The amount of insurance not used up is $700 ($1,200 minus $500) and should show up on the Balance Sheet as Prepaid Expense. If the Farmers had no Prepaid Expense at the beginning of the year, they would calculate Change in Prepaid Insurance as follows:

Amount of the Prepaid Insurance not yet used up at the end of the year	$700
− Prepaid Expense (insurance portion) at the beginning of the year	− 0
= Change in Prepaid Insurance	$700

If the Farmers recorded journal entries to make adjustments, they would have recorded the following journal entry and posted it to the ledger. The financial statements would be accrual-adjusted when they were prepared.

(37)	Dec. 31	1300 Prepaid Expenses	700	
		6820 Change in Prepaid Insurance		700

This adjustment reports the proper amount of insurance expense on the Income Statement. The $700 Change in Prepaid Insurance offsets the $1,200 payment recorded in the Insurance account. Therefore, the records indicate that $1200 was paid for insurance but $700 of that amount has not been used up. The net amount of Insurance and Change in Prepaid Insurance is $500, which is the actual expense for Insurance.

Income Statement:
    Operating Expenses:

Insurance	($1,200)
+ Change in Prepaid Insurance	700
Net Effect on Net Farm Income from Operations	($500)

Balance Sheet:
    Assets:

Prepaid Expenses	$700

The journal entries in Table 4-8 demonstrate the difference between cash-basis accounting and the accrual-adjusted approach for prepaid insurance.

The cash-basis system does not report any prepaid expenses. The accrual-adjusted approach provides a more accurate report of the farm's assets by reporting the amount of unused insurance and the correct amount of the expense for the insurance coverage used up as the net amount of Insurance and Change in Prepaid Insurance. These procedures can be used for other types of prepaid expenses, such as rent or any other expenditure that results in some other party owing goods or services to the farm business.

Cash Investment in Growing Crops

A special kind of prepaid expense is Cash Investment in Growing Crops for perennial crops. The accounting procedures are similar to other prepaid expenses. The farm accountant records the costs of growing the crop (such as cultivation, pruning,

TABLE 4-8 ■ *Cash-Basis vs Accrual-Adjusted (prepaid insurance).*

Cash-Basis			Accrual-Adjusted		
Purchase			Purchase		
Insurance	1,200		Insurance	1,200	
Cash		1,200	Cash		1,200
Adjustment when financial statements are prepared			Adjustment when financial statements are prepared		
(none)			Prepaid Expenses	700	
			Change in Prepaid Insurance		700

fertilizing, plant depreciation, and so on) as various operating expenses as they occur. If the harvest of the crop occurs in the year following the year that it was produced, the balance sheet should report the costs of growing the crop as a prepaid expense, such as Cash Investment in Growing Crops. These costs are considered a prepaid expense because the costs have been paid for in advance of the year of the sale. The contra or adjunct account on the income statement is **Change in Investment in Growing Crops**, calculated as the difference between the growing costs for this year and the balance in Cash Investment in Growing Crops at the end of last year. The effect on the income statement is similar to the effect of prepaid expenses: Operating Expenses +/− Change in Investment in Growing Crops = accrual–adjusted operating expenses.

■ Determine the amount of costs incurred for the perennial crop not harvested at the end of the year.
   • Make adjustment:
      • Balance sheet shows amount of the costs incurred
         • Cash Investment in Growing Crops
      • Income statement adjusts the operating expenses
         • (Operating Expenses)
         • +/− Change in Change in Investment in Growing Crops

Steve and Chris had developed an apple orchard and recorded the development in journal entry (14). In the year of their first crop of apples, they recorded various operating expenses for growing the apples that amounted to $2,500. If they do not harvest the apple crop before the financial statements are prepared at the end of the year, they need to make adjustments to the financial statements to report a prepaid expense (Cash Investment in Growing Crop) on the Balance Sheet for $2,500 and to make an adjustment to operating expenses on the Income Statement (Change in Investment in Growing Crops). They did not report any Cash Investment in Growing Crops at the beginning of the year, so they calculate the Change in Investment in Growing Crops as follows:

Amount of the Cash Investment in Growing Crops at the end of the year	$2,500
Cash Investment in Growing Crops at the beginning of the year	− 0
= Change in Investment in Growing Crops	$2,500

If the Farmers recorded journal entries to make adjustments, they would have recorded the following journal entry and posted it to the ledger. The financial statements would be accrual-adjusted when they were prepared.

(38)   Dec. 31   1400 Cash Investment in Growing Crops   2,500
                          6830 Change in Investment in Growing Crops   2,500

Instead of reporting expenses associated with the perennial crop, the expenditures are reported as a prepaid expense (an asset), the benefits of which will be reaped the following year when the crop is harvested.

Income Statement:
                    Various operating expenses for perennial crop   ($2,500)
                    + Change in Investment in Growing Crops $2,500
                            Net Effect on Net Farm Income from Operations $ 0

Balance Sheet:
        Assets:
                    Cash Investment in Growing Crops $2,500

The journal entries in Table 4-9 demonstrate the difference between cash-basis accounting and the accrual-adjusted approach for cash investments in growing crops.

The cash-basis system reports the growing costs as expenses. The FFSC recommends that costs of growing perennial crops be reported as an investment; therefore, a prepaid expense is reported on the balance sheet instead of expenses on the income statement in the accrual-adjusted approach.

PRACTICE WHAT YOU HAVE LEARNED   *Practice what you have learned about adjustments and complete Problem 4-4 at the end of the chapter.*

Depreciation expense

The purchase of non-current assets (breeding livestock, machinery and equipment, perennial crops, buildings and improvements, and leased assets) is another type of prepaid expense. Unlike feed or insurance, non-current assets last longer than a few

TABLE 4-9 ■ *Cash-Basis vs. Accrual-Adjusted (cash investment in growing crops).*

Cash-Basis		Accrual-Adjusted	
When growing expenses occur:		When growing expenses occur:	
Operating expenses 2,500		Operating expenses 2,500	
Cash 2,500		Cash 2,500	
Adjustment when financial statements are prepared		Adjustment when financial statements are prepared	
(none)		Cash Investment in Growing Crops 2,500	
		Change in Investment in 2,500	
		Growing Crops	

months or a year. **Depreciation expense** is the annual allocation of the cost of non-current assets (except land) over the expected length of time that each asset is expected to be used in the farm operation (the estimated useful life). The amount of depreciation for each asset represents the amount of the asset "used up." Unlike feed, however, the amount used up is not an amount that has been literally "used up"; rather it represents a using up of the asset each year. Each year that an asset is used, a certain amount of that use is reported as Depreciation Expense. The use of the estimated life of the asset coincides with the going concern concept mentioned in Chapter 1 because the farm is expected to operate indefinitely. Some assets, like buildings, might outlive the current owners, but the farm business will continue to operate, so the building's expected useful life is used in the calculation of Depreciation Expense. You will learn about how to calculate depreciation expense in Chapter 6.

Suppose that the Farmers determine that depreciation expense for 20X1 is $22,150. If the Farmers recorded journal entries to make adjustments, they recorded the following journal entry and posted it to the ledger. The financial statements would be accrual-adjusted for depreciation when they were prepared.

(39)	Dec. 31	6780 Depreciation Expense	22,150	
		1980 Accumulated Depreciation		22,150

Income Statement:
Depreciation Expense $22,150

Balance Sheet:
    Assets:
        Non-Current Assets:

Office Furniture	$ 1,000
Buildings, Improvements	110,000
Tractor	50,000
Truck	15,000
Leased Harvester	100,000
Orchard	45,000
Accumulated Depreciation	(22,150)

Exercise 4-3 *Think about the assets for your 4-H project. Which ones would be depreciated? How long is the estimated useful life of each? Answer: You would depreciate the clipping chute and perhaps the clippers. Deciding on which assets should be depreciated depends on how long they will last and how much the assets cost. Some assets such as grooming supplies (brushes, combs, and so on) might last more than one year, but the amount of depreciation would be so small that it would not affect the income statement enough to make the extra bookkeeping worthwhile. The clipping chute is a rather large expenditure for a 4-H project and the clippers also might be. If the cost of the clippers is not significant enough to depreciate, then you can record the cost of the clippers as a supply expense. The estimated life of the chute depends on how many years that you expect to be showing cattle.*

Accrued Expenses

Learning Objective 5 ■ To perform adjustments for accrued expenses.

Accrued expenses refer to expenses that have occurred for the farm business but have not yet been paid for. Certain expenses "build up" (hence the term "accrued") but are not yet paid for, either because they are not yet due or because the cash is not yet available. The farm accountant needs to determine whether any accrued expenses exist at the time that financial statements are prepared. Because these expenses have not yet been paid, they have not been recorded, but they are costs of the farm business that have occurred for the year. An adjustment is required at the end of the year (or whenever financial statements are being prepared) to record the appropriate amount of expense for the year.

- ■ Determine the amount of unpaid expenses that have accrued by the end of the year.
  - Make adjustment:
    - Balance sheet shows amount of unpaid expense
    - Income statement shows amount of accrued expense

Accrued interest

A primary example of accrued expense is interest that is accruing on all of the farm loans. Even if loan and interest payments are made throughout the year, by the end of the year, interest has accrued since the last payment. Under the cash-basis system, interest expense is recorded only when it is paid. The accrual-adjusted approach also reports the accrued interest. To make the adjustments, the farm accountant needs to calculate the amount of accrued interest, and must keep accurate records of principal amounts borrowed, interest rates, the dates on which the loans were borrowed, and when the last payments were made. The balance sheet reports the amount of accrued interest as Interest Payable. The adjustment on the income statement is Change in Interest Payable. **Change in Interest Payable** is a contra or adjunct account presented with Interest Expense to report the correct amount of interest for the year.

- ■ Determine the amount of interest that has accrued by the end of the year.
  - Make adjustment:
    - Balance sheet shows amount of accrued interest
      - Interest Payable
    - Income statement shows amount of interest for the year
      - (Interest Expense)
      - +/− Change in Interest Payable

In Chapter 3, Steve and Chris borrowed $50,000 to purchase a tractor on March 1, 20X1 in journal entry (7). They will make one payment a year later and pay off the entire note. No payments for interest or principal will be made until March 1 next year. At the end of 20X1, the Farmers calculate the amount of the accrued interest on this loan. Suppose that the amount of accrued interest is $5,000. This amount, $5,000, is the amount of accrued interest that should be reported on the Balance Sheet as Interest Payable for 20X1. If the

Farmers reported no Interest Payable at the beginning of the year, they would calculate the Change in Interest Payable as follows:

Amount of the interest accrued at the end of the year	$5,000
− Interest Payable at the beginning of the year	− 0
= Change in Interest Payable	$5,000

If the Farmers recorded journal entries to make adjustments, they would have recorded the following journal entry and posted it to the ledger. The financial statements would be accrual-adjusted when they were prepared.

(40)	Dec. 31	8110 Change in Interest Payable	5,000	
		2200 Interest Payable		5,000

The Income Statement reflects the amount of interest expense that occurred in 20X1, including the amount that is not due until March 1, 20X2. Steve and Chris had made a payment for interest for a different loan on October 1, 20X1, for $1,200 in journal entry (8) in Chapter 3, so Interest Expense has a balance of $1,200.

Income Statement:

Interest Expense	($1,200)
− Change in Interest Payable	(5,000)
Net Effect on Net Farm Income from Operations	($6,200)

Balance Sheet:
Liabilities:

Interest Payable	$5,000

The journal entries in Table 4-10 demonstrate the difference between cash-basis accounting and the accrual-adjusted approach for accrued interest.

The cash-basis system reports neither the amounts owed for interest on outstanding loans nor the correct amount of interest expense. The accrual-adjusted approach adjusts the income statement for the correct amount of interest expense with the Change in Interest Payable, and also reports a more accurate amount of liabilities by reporting Interest Payable for the amount of unpaid interest.

Unpaid bills

As bills are paid throughout the year, the farm accountant records the amounts in the appropriate expense accounts. However, if the farm business has unpaid bills at the end of the year, expenses have occurred and these costs need to be reported on accrual-adjusted farm financial statements. At the end of the year (or when financial

TABLE 4-10 ■ *Cash-Basis vs. Accrual-Adjusted (interest owed).*

Cash-Basis	Accrual-Adjusted	
Adjustment when financial statements are prepared	Adjustment when financial statements are prepared	
(none)	Change in Interest Payable	5,000
	Interest Payable	5,000

statements need to be prepared), the farm accountant needs to determine the total of the unpaid bills. In the accrual-adjusted approach, the balance sheet reports an adjustment for these unpaid bills as Accounts Payable. The adjustment on the income statement is **Change in Accounts Payable**, a contra or adjunct account reported with operating expenses to reflect the correct amount of expenses that have occurred for the year.

- Determine the amount of unpaid bills at the end of the year.
  - Make adjustment:
    - Balance sheet shows amount of unpaid bills
      - Accounts Payable
    - Income statement shows amount of the bills for the year
      - (Operating Expenses)
      - +/– Change in Accounts Payable

Suppose that the amount of unpaid bills for the Farmers at the end of 20X1 is $340. If Accounts Payable had a zero balance at the beginning of the year, the Change in Accounts Payable is computed as follows:

Amount of the bills owed at the end of the year	$340
– Accounts Payable at the beginning of the year	– 0
= Change in Accounts Payable	$340

If the Farmers recorded journal entries to make adjustments, they would have recorded the following journal entry and posted it to the ledger. The financial statements would be accrual-adjusted when they were prepared.

(41)	Dec. 31	6810 Change in Accounts Payable	340	
		2000 Accounts Payable		340

The income statement reflects the correct amount of operating expenses that occurred in 20X1, including the amount that was not paid yet.

Income Statement:

Operating Expenses	$XXXX
– Change in Accounts Payable	(340)
Net Effect on Net Farm Income from Operations	($340)

Balance Sheet:
Liabilities:

Accounts Payable	$340

The journal entries in Table 4-11 demonstrate the difference between cash-basis accounting and the accrual-adjusted approach for unpaid bills.

The cash-basis system reports neither the amounts owed for the unpaid bills nor the correct amount of expenses. With the Change in Accounts Payable, the accrual-adjusted approach adjusts the operating expenses to the correct amount of expenses that have occurred. The balance sheet reports a more accurate amount of liabilities by reporting Accounts Payable for unpaid bills.

TABLE 4-11 ■ *Cash-Basis vs. Accrual-Adjusted (unpaid bills).*

Cash-Basis	Accrual-Adjusted	
Adjustment when financial statements are prepared	Adjustment when financial statements are prepared	
(none)	Change in Accounts Payable	340
	Accounts Payable	340

Unpaid taxes

The farm accountant records the amount of taxes paid in Payroll Taxes, Real Estate and Property Taxes, or other tax expense accounts. If the farm operation owes taxes at the end of the year, the farm accountant makes similar adjustments for the unpaid taxes. The balance sheet reports the amount of taxes owed as Taxes Payable. The adjustment on the income statement is Change in Taxes Payable. **Change in Taxes Payable** is a contra or adjunct account reported with Income Tax Expense to report the correct amount of taxes for the year.

> **PRACTICE WHAT YOU HAVE LEARNED** *Practice what you have learned about adjustments for accrued expenses by completing Problem 4-5 at the end of the chapter.*

Accrued Revenues

Accrued revenues refer to earned income for which the cash has not yet been received. In the cash-basis system, revenues are recorded only when cash is received. Under the accrual-adjusted approach, the farm accountant records all amounts of money earned, including any money that has not been received yet. Revenues are earned when a sale has been made and the product has been delivered. An example of accrued revenues is delivering grain to the local grain elevator but not being paid for the grain until some time in the future. At the end of the year, the farm accountant must determine if any money is owed to the farm business from the sale of farm products.

Learning Objective 6 ■ To perform adjustments for accrued revenues.

Accounts Receivable

The balance sheet reports an adjustment for accrued revenues as Accounts Receivable. The adjustment on the income statement is Change in Accounts Receivable. **Change in Accounts Receivable** is a contra or adjunct account reported with cash revenues to reflect the correct amount of money earned. An increase in Change in Accounts Receivable represents an increase in revenue for the farm business and is included in the computation of Gross Revenue.

- ▪ Determine the amount of money owed to the farm business at the end of the year.
  - Make adjustment:
    - Balance sheet shows amount of accrued revenue
      - Accounts Receivable
    - Income statement shows amount of revenue earned
      - Cash Sales (from crops, feed, and/or market livestock)
      - +/− Change in Accounts Receivable

Steve and Chris determined that $3,500 was owed to them at the end of 20X1. The accrual-adjusted system requires that the Balance Sheet reports the amount owed to them as Accounts Receivable. They adjust the Income Statement for Change in Accounts Receivable. If Accounts Receivable had a zero balance at the beginning of the year, they calculate the Change in Accounts Receivable as follows:

Amount of money owed to the farm business at the end of the year	$3,500
− Accounts Receivable at beginning of the year	− 0
= Change in Accounts Receivable	$3,500

If the Farmers recorded journal entries to make adjustments, they would have recorded the following journal entry and posted it to the ledger. The financial statements would be accrual-adjusted when they were prepared.

(42)	Dec. 31	1100 Accounts Receivable	3,500	
		4700 Change in Accounts Receivable		3,500

The Income Statement reflects the amount of revenue earned in 20X1, including the amount of money that was not yet received.

Income Statement:

Cash Sales of Market Livestock and Poultry	$XXX
+ Change in Accounts Receivable	3,500
Effect on Gross Revenue	$3,500

Balance Sheet:
  Assets:

Accounts Receivable	3,500

The journal entries in Table 4-12 demonstrate the difference between cash-basis accounting and the accrual-adjusted approach for accrued revenues.

The cash-basis system reports neither the amounts owed to the farm business for the sale of the farm products nor the correct amount of revenue. The accrual-adjusted approach adjusts the gross revenue to the correct amount for the revenue that has been earned as the Change in Accounts Receivable and also reports a more accurate amount of assets by reporting Accounts Receivable for the money owed to the farm business.

TABLE 4-12 ■ *Cash-Basis vs. Accrual-Adjusted (accrued revenues).*

Cash-Basis	Accrual-Adjusted	
Sale	Sale	
(none)	(none)	
Adjustment when financial statements are prepared	Adjustment when financial statements are prepared	
(none)	Accounts Receivable	3,500
	Change in Accounts Receivable	3,500

Change in value of raised breeding livestock due to age progression

The cost of raising breeding livestock is sometimes difficult to determine because the record keeping for accumulating and determining such costs is complex. The FFSC recommends that when the cost of raising breeding livestock is not easily determined, base values can be assigned for various age groups of breeding livestock. We use base values to estimate the cost of producing the breeding stock and the value of the animals at different points in their productive lives. The farm accountant can obtain base values from reputable sources that report the approximate cost of raising the livestock.

Breeding livestock move through age groups each successive year after they are put into production. The values of the animals change as the animals proceed through each of the different age groups while they remain in the herd or flock. The result is a change in the value of breeding stock due to age progression. In the FFSC Guidelines, this change in value is reported as **Change in Value Due to Change in Quantity of Raised Breeding Livestock**. As the productive value of the animal increases, the value to the farm business increases. The farm accountant makes an adjustment at the end of the year to report these changes in value.

Learning Objective 7 ■ To record changes in value for raised breeding livestock.

- Determine the amount of changes in value for the animals that moved to the next age group.
  - Make adjustment:
    - Balance sheet shows new value for raised breeding livestock
      - Breeding Livestock (adjusted for age progression)
    - Income statement shows effect on Gross Revenue
      - +/– Change in Value Due to Change in Quantity of Raised Breeding Livestock

Steve and Chris owned a cow herd at the beginning of the year 20X1. By the end of the year, some of the animals have moved into the next age group, so their base values have changed. If Steve and Chris determined that the change in the value of their raised cattle herd due to age progression was an increase of $6,000, they would report an adjustment on the Balance Sheet by adding $6,000 to the balance for Breeding Livestock.

If the Farmers recorded journal entries to make adjustments, they would have recorded the following journal entry and posted it to the ledger. The financial statements would be accrual-adjusted for changes in value due to age progression when they were prepared.

| (43) | Dec. 31 | 1500 Breeding Livestock | 6,000 | |
| | | 4600 Change in Value Due to Change in Quantity of Raised Breeding Livestock | | 6,000 |

If they had determined that the change in value was a decrease, the entry would be reversed, with the Change in Value account being debited and the Breeding Livestock account being credited. The Income Statement reports this increase in value as part of Gross Revenue.

Income Statement:

+ Change in Value Due to Change in Quantity of Raised Breeding Livestock	$6,000	
Effect on Gross Revenue	$6,000	

Ledger:

1500 Breeding Livestock

Date	Description	JE#	Debits	Credits	Balance
Beginning Balance					0
Jan. 2	Set up farm account for breeding cattle	(1)	76,000		76,000
Mar. 10	Sale of bull to neighbor	(10)		1,000	75,000
Nov. 20	Sale of culled cows	(19)		1,500	73,500
Dec. 31	Adjustment	(43)	6,000		79,500

Balance Sheet:
Assets:
Breeding Livestock $79,500

Changes in base values of raised breeding livestock

Base values are assigned for each age group, and from time to time these base values may change due to new estimates of the cost of raising and maintaining the animals. The preceding adjustment using Change in Value Due to Change in Quantity of Raised Breeding Livestock assumes that base values are unchanged (see Figure 4-5).

If the base values change, the farm accountant makes an additional adjustment to reflect the change in the value of raised breeding livestock due to the change in the base values. This adjustment is reported as **Gains/Losses Due to Changes in General Base Values of Breeding Livestock** on the income statement. An adjustment is also made to Breeding Livestock for any changes in base values.

FIGURE 4-5 ■ *Changes in Base Values vs. Changes Due to Age Progression.*

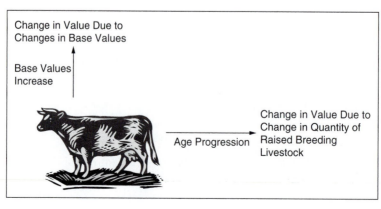

- ▪ Determine the amount of changes in base values for the entire herd.
  - Make adjustment:
    - Balance sheet shows new value for raised breeding livestock
      - Breeding Livestock (adjusted for changes in base values)
    - Income statement shows effect on accrual adjusted net income
      - +/− Gains/Losses Due to Changes in General Base Values of Breeding Livestock

Suppose that the Farmers determined an increase in base values of $4,500. If the Farmers recorded journal entries to make adjustments, they would have recorded the following journal entry and posted it to the ledger. The financial statements would be accrual-adjusted for changes in base values when they were prepared.

Dec. 31	1500 Breeding Livestock	4,500
	8300 Gains/Losses Due to Changes in General Base Values of Breeding Livestock	4,500

The Income Statement reports the change in value as Gains/Losses Due to Changes in General Base Values of Breeding Livestock but it is not part of Gross Revenue.

Gains/Losses Due to Changes in General Base Values of Breeding Livestock	$4,500
Effect on Accrual Adjusted Net Farm Income	$4,500

An increase in base values is recorded as a credit to the account and reported as a gain. If the base values had declined, the entry would be recorded as a debit and would be reported as a loss instead of a gain.

Income Taxes

Income tax expense

Income taxes paid by a farm operation are based on cash-basis income. The actual amount of taxes paid will also depend upon the tax laws and the specific circumstances of the farm business. Accrual-adjusted financial statements report an estimated amount of taxes owed as Taxes Payable on the balance sheet at the end of the year. An adjustment is also made on the income statement as **Change in Taxes Payable**, a contra or adjunct account that includes the difference between the amount of taxes owed at the end of the previous year and the amount of taxes owed at the end of the current year.

Learning Objective 8 ▪ To perform adjustments for income taxes, current deferred taxes, and the first component of non-current deferred taxes.

- ▪ Estimate the amount of income tax for the year.
  - Make adjustment:
    - Balance sheet shows amount of taxes owed
      - Taxes Payable
    - Income statement shows amount of income tax for the year
      - (Income Tax Expense)
      - +/− Change in Taxes Payable (for difference in Taxes Payable from last year)

When the Farmers began their farm business in 20X1, they owed $2,160 for farm income tax at the end of the previous year. In journal entry (2) in Chapter 3, they correctly recorded this liability as Taxes Payable. After consulting with their local certified public accountant, they estimate that they will owe $3,030 in income tax for the year 20X1 (which they will pay in 20X2). The Farmers adjust the Balance Sheet so that the reported tax liability (Taxes Payable) is $3,030. They also make an adjustment on the Income Statement as Change in Taxes Payable for the difference between the amount that they owed at the end of 20X0 ($2,160) and the amount that they owe at the end of 20X1 ($3,030). They calculate Change in Taxes Payable as follows:

Amount of the taxes owed at the end of the year	$3,030
− Taxes Payable at the beginning of the year	− 2,160
= Change in Taxes Payable	$ 870

If the Farmers recorded journal entries to make adjustments, they would have recorded the following journal entry and posted it to the ledger. The financial statements would be accrual-adjusted for the difference between the income tax liability from last year and the income tax liability for the current year:

(44)	Dec. 31	9110 Change in Taxes Payable	870	
		2100 Taxes Payable		870

Income Tax Expense reports the amount of 20X0 income taxes that the Farmers paid in 20X1. The Change in Taxes Payable is an expense item that adjusts the amount of taxes paid for the additional amount of taxes owed for 20X1. The amount of $3,030 is the total accrual-adjusted income tax expense for the year 20X1.

Income Statement:

Income before Taxes	$XXX
Income Tax Expense	(2,160)
− Change in Taxes Payable	(870)
Net Effect on Accrual Adjusted Net Income	($3,030)

Ledger:

2100 Taxes Payable

Date	Description	JE#	Debits	Credits	Balance
Beginning Balance					0
Jan. 2	To set up farm taxes payable	(2)		2,160	2,160
Apr. 1	To record payroll withholdings	(26)		182	2,342
Apr. 10	Payment of payroll withholdings	(27)	182		2,160
Dec. 31	Adjustment	(44)		870	3,030

Balance Sheet:
Liabilities:

Taxes Payable	$3,030

The journal entries in Table 4-13 demonstrate the difference between cash-basis accounting and the accrual-adjusted approach for the income taxes owed.

The cash-basis system reports only the amount of income taxes paid. However, the income taxes paid pertain to the previous calendar year, not the current year.

TABLE 4-13 ▪ *Cash-Basis vs. Accrual-Adjusted (income taxes).*

Cash-Basis			Accrual-Adjusted		
When income taxes are paid			When income taxes are paid		
Income Tax Expense	2,160		Income Tax Expense	2,160	
Cash		2,160	Cash		2,160
Adjustment when financial statements are prepared			Adjustment when financial statements are prepared		
(none)			Change in Taxes Payable	870	
			Taxes Payable		870

The current year's income taxes should be reported in the current year along with the current year's income. Under the accrual-adjusted approach, the adjustment at the end of the year approximates the income tax liability for the current year's income.

Current deferred taxes

Cash-basis income is usually different from accrual-adjusted net income. To illustrate the potential difference in income between these two systems, Table 4-14 presents sample income statements. The cash-basis income statement is based on the journal entries (8) through (31) in Chapter 3. The accrual-adjusted income statement is based on the journal entries (8) through (31) from Chapter 3 and (32) through (44), except (36), presented in this chapter.

    The income tax liability for the year 20X1 ($3,030) is based on net cash income of $52,036 and other provisions of current tax laws (including tax depreciation). The Taxes Payable account had a balance of $2,160 from journal entry (2). Journal entry (44) adjusts that balance up to $3,030. This adjustment is based on cash-basis net income. However, this amount does not reflect income tax expense based on the accrual-adjusted net income, which is a more accurate picture of the financial performance of the farm business (see Figure 4-6). When farm financial statements are prepared according to the accrual-adjusted approach, the income taxes should be based on accrual-adjusted net income, even though the actual amount of income taxes paid is based primarily on cash-basis numbers. The income taxes associated with accrual-adjusted net income should be reported in the same year that the accrual-adjusted net income was earned.

    The difference in the tax liabilities between the cash-basis income and the accrual-adjusted income is an example of income taxes that will eventually be resolved in future years. These taxes are known as **deferred tax liabilities**. They represent tax liabilities that will be paid in the future when the cash transactions are completed. The deferred taxes that will be resolved the next year are reported as **Current Deferred Taxes**. (The method to calculate the amount of these taxes is discussed in Chapter 9.) The following items result in current deferred taxes:

- The timing of cash transactions between cash-basis net income and accrual-adjusted net income. These timing differences are reflected in the "change items" that were recorded in journal entries (32) through (42), except (36).

TABLE 4-14 ■ *Cash-Basis versus Accrual-Adjusted Income Statements.*

Cash-Basis Income Statement			Accrual-Adjusted Income Statement		
Revenue:			Revenue:		
Cash Crop Sales	$12,600		Cash Crop Sales	$12,600	
			Change in Crop Inventory	8,300	
Sales of Market Livestock	50,900		Sales of Market Livestock	50,900	
Loss from Sale of Culled Livestock	(250)		Loss from Sale of Culled Livestock	(250)	
			Change in Quantity of		
			Raised Breeding Livestock	6,000	
			Change in Accounts Receivable	3,500	
Total Revenue		$63,250	Gross Revenue		$81,050
Expenses:			Expenses:		
			Change in Investment		
			in Growing Crops	2,500	
Feeder Livestock	(1,500)		Feeder Livestock	(1,500)	
			Change in Purchased Feeder		
			Livestock Inventory	750	
Purchased Feed	(1,000)		Purchased Feed	(1,000)	
			Change in Purchased Feed Inventory	230	
Wage Expense	(1,200)		Wage Expense	(1,200)	
Payroll Tax Expense	(129)		Payroll Tax Expense	(129)	
Truck and Machine Hire	(150)		Truck and Machine Hire	(150)	
Herbicides, Pesticides	(500)		Herbicides, Pesticides	(500)	
Livestock Supplies	(75)		Livestock Supplies	(75)	
Insurance	(1,200)		Insurance	(1,200)	
			Change in Prepaid Insurance	700	
Real Estate/Property Taxes	(1,300)		Real Estate/Property Taxes	(1,300)	
			Change in Accounts Payable	(340)	
			Depreciation Expense	(22,150)	
Interest Expense	(1,200)		Interest Expense	(1,200)	
			Change in Interest Payable	(5,000)	
Net Farm Income from Operations		54,996	Net Farm Income from Operations		49,486
Loss on Sale of Capital Assets	(800)		Loss on Sale of Capital Assets	(800)	
Income before Taxes		54,196	Income before Taxes		48,686
Income Tax Expense	(2,160)		Income Tax Expense	(2,160)	
			Change in Taxes Payable	(870)	
Net Cash Income (Loss)		$52,036	Accrual Adjusted Net Income		$45,656

FIGURE 4-6 ■ *Income and Income Tax Liabilities.*

```
Income                          Income Tax Liabilities
                                Income Tax Expense      $2,160
                                Change in Taxes Payable    870
Net Cash Income      $52,036 ───────────────────────▶  $3,030

+/− Change Items*

= Accrual Adjusted Net Income  $40,906 ────────────▶  ??????

* See journal entries (32) through (44), except (36).
```

The balance sheet reports the current deferred taxes. The income statement presents an adjustment for the difference between last year's amount of current deferred taxes and this year's amount of current deferred taxes.

- ▪ Estimate the amount of current deferred taxes for the year.
  - • Make adjustment:
    - • Balance sheet shows amount of current deferred taxes
      - • Current Deferred Taxes
    - • Income statement shows amount of income tax for the year
      - • (Income Tax Expense)
      - • +/– Change in Taxes Payable (for difference in Taxes Payable from last year)
      - • +/– Change in Taxes Payable (for difference in Current Deferred Taxes from last year)

If the Farmers determine that the amount of current deferred taxes at the end of 20X1 is $1,450, they should adjust the Balance Sheet to report $1,450 for Current Deferred Taxes. The adjustment on the Income Statement is part of Change in Taxes Payable. The Farmers had no current deferred taxes at the beginning of the year, so they calculate Change in Taxes Payable pertaining to current deferred taxes as follows:

Amount of the current deferred taxes owed at the end of the year	$1,450
– Current Deferred Taxes at the beginning of the year	– 0
= Change in Taxes Payable	$1,450

If the Farmers recorded journal entries to make adjustments, they would have recorded the following journal entry and posted it to the ledger. The financial statements would be accrual-adjusted for current deferred taxes when they were prepared.

(45)	Dec. 31	9110 Change in Taxes Payable	1,450	
		2510 Current Deferred Taxes		1,450

Because of this adjustment, the Income Statement reports the appropriate amount of income tax expense in the year in which the accrual-adjusted income was earned.

Income Statement:

Income before Taxes	$XXX
Income Tax Expense	(2,160)
+ Change in Taxes Payable	(870)
+ Change in Taxes Payable	(1,450)
Effect on Accrual Adjusted Net Income	($4,480)

Balance Sheet:
Liabilities:

Taxes Payable	$3,030
Current Deferred Taxes	1,450

The journal entries in Table 4-15 demonstrate the difference between cash-basis accounting and the accrual-adjusted approach for current deferred taxes.

The cash-basis system reports neither the amounts owed by the farm business for the deferred taxes nor the correct amount of income tax expense. The accrual-adjusted

TABLE 4-15 ■ *Cash-Basis vs. Accrual-Adjusted (current deferred taxes).*

Cash-Basis	Accrual-Adjusted	
Adjustment when financial statements are prepared	Adjustment when financial statements are prepared	
(none)	Change in Taxes Payable	1,450
	Current Deferred Taxes	1,450

approach adjusts the income tax expense to the correct amount with the Change in Taxes Payable and reports a more accurate amount of liabilities by reporting Current Deferred Taxes.

Non-current deferred taxes, first component

Non-current deferred taxes arise from the differences between the book or base values of non-current assets and their tax basis, and also from the differences between the market values of non-current assets and their book or base values. (Examples are provided in Chapter 9.) You will recall the definition and computation of book value.

Book Value = Original cost of an asset − Accumulated financial depreciation of the asset

The accumulated depreciation in this definition is based on depreciation methods for financial accounting, such as the straight-line method illustrated in journal entry (39). Tax rules allow for different methods of calculating depreciation. The **tax basis** refers to an alternative computation for book value using tax depreciation methods for the accumulated depreciation.

Tax Basis = Original cost of an asset − Accumulated tax depreciation of the asset

The difference between the tax depreciation and the financial depreciation results in deferred taxes. (More on this in Chapter 9.) The balance sheet reports these tax liabilities as **Non-Current Deferred Taxes** because they are not likely to be resolved during the next year.

A non-current deferred tax consists of two components. The first component relates to

■ the difference between base values and tax basis of raised breeding livestock and
■ the difference between book values and tax basis of other non-current assets

The balance sheet reports the deferred taxes for the first component as part of the non-current deferred tax liabilities. The adjustment on the income statement for the first component is part of Change in Taxes Payable.

■ Estimate the amount of the first component of non-current deferred taxes for the year.
  • Make adjustment:
    • Balance sheet shows amount of non-current deferred taxes
      • Non-current Deferred Taxes

- Income statement shows amount of income tax for the year
  - (Income Tax Expense)
  - +/− Change in Taxes Payable (for difference in Taxes Payable from last year)
  - +/− Change in Taxes Payable (for difference in Current Deferred Taxes from last year)
  - +/− Change in Taxes Payable (for difference in first component of Non-current Deferred Taxes from last year)

Suppose that the Farmers estimate that the amount of the first component of non-current deferred taxes current is $3,300. They had no non-current deferred taxes pertaining to the first component at the beginning of the year, so they calculate the Change in Taxes Payable pertaining to the first component as follows:

Amount of the first component of non-current deferred taxes at the end of the year	$3,300
− Non-Current Deferred Taxes (first component) at the beginning of the year	− 0
= Change in Taxes Payable	$3,300

If the Farmers recorded journal entries to make adjustments, they would have recorded the following journal entry and posted it to the ledger. The financial statements would be accrual-adjusted for the first component of non-current deferred taxes when they were prepared.

(46)	Dec. 31	9110 Change in Taxes Payable	3,300	
		2500 Non-Current Deferred Taxes		3,300

The Income Statement reports this adjustment as part of Change in Taxes Payable.

Income Statement:

Income before Taxes	$XXX
Income Tax Expense	(2,160)
+ Change in Taxes Payable	(870)
+ Change in Taxes Payable	(1,450)
+ Change in Taxes Payable	(3,300)
Effect on Accrual Adjusted Net Income	($7,780)

Balance Sheet:
Liabilities:
Current Liabilities:

Taxes Payable	$3,030
Current Deferred Taxes	1,450
Non-Current Liabilities:	
Non-Current Deferred Taxes	3,300

The journal entries in Table 4-16 demonstrate the difference between cash-basis accounting and the accrual-adjusted approach for the first component of non-current deferred taxes.

The cash-basis system reports neither the amounts owed by the farm business for the non-current deferred taxes nor the correct amount of income tax expense.

TABLE 4-16 ■ *Cash-Basis vs. Accrual-Adjusted (first component of non-current deferred taxes).*

Cash-Basis	Accrual-Adjusted	
Adjustment when financial statements are prepared	Adjustment when financial statements are prepared	
(none)	Change in Taxes Payable	3,300
	Non-Current Deferred Taxes	3,300

The accrual-adjusted approach adjusts the income tax expense to the correct amount with the Change in Taxes Payable and reports a more accurate amount of liabilities by reporting Non-Current Deferred Taxes. Chapter 5 presents a discussion of the second component of non-current deferred taxes.

> **PRACTICE WHAT YOU HAVE LEARNED** *Test your knowledge of accounting for income taxes by completing Problem 4-7 at the end of the chapter.*

POSTING

Learning Objective 9 ■ To post adjusting journal entries.

If the farm accountant chooses to record journal entries to make adjustments as illustrated in this chapter, the next step in the year-end procedures is to post them to the ledger accounts. The posting procedures are identical to those for the transactions presented in Chapter 3. The debits and credits from each of the journal entries are copied to each of the corresponding ledger accounts.

> **Exercise 4-4** *Appendix F contains the Farmers' ledger accounts after posting the adjusting journal entries in Chapter 4. Review the adjusting entries (32) through (46), except (36). Verify that each of the adjusting journal entries have been posted correctly (that is, check that each debit and each credit have been copied correctly to the appropriate account).*

> **PRACTICE WHAT YOU HAVE LEARNED** *Practice the posting procedures by completing Problem 4-8 at the end of the chapter.*

TRIAL BALANCE

Learning Objective 10 ■ To prepare an adjusted trial balance.

The **trial balance** is a list of the balances of all accounts in the ledger. The trial balance serves as the basis for preparing the financial statements. The purpose of the trial balance is to check that the sum of all debits equals the sum of all credits. If they do not, then an error has occurred. Errors such as incorrect posting of journal entries or calculation of account balances can occur in a manual accounting system. The farm accountant prepares a trial balance before preparing the financial statements to correct any errors so that these errors do not carry over to the financial statements. If the debits equal the credits, the farm accountant can be assured that the errors indicated above have not occurred. However, if some transactions have not been recorded, or if a journal entry was not posted or was posted twice, or the amounts in journal entry are incorrect, the trial balance will not reflect these errors and the farm accountant must be aware that these other types of errors can occur. If the farm accountant records and posts the journal entries to make adjustments, the trial balance is called an **Adjusted Trial Balance**, and it includes all of the accounts in which adjustments were made. A trial balance without the accounts used to make adjustments is called an **Unadjusted Trial Balance**, like the trial balance in Appendix E.

Before preparing the trial balance, the farm accountant must calculate the account balances. Then, all of the accounts are listed in the order in which they appear in the ledger with one column for the debit balances and one column for the credit balances, creating the trial balance. To know which balances are debits and which are credits, the farm accountant must recall which column records the increases for each type of account.

- Increases are recorded as debits for assets, expenses, and owner withdrawals.
- Increases are recorded as credits for liabilities, equity, and revenue accounts.

As you learned in Chapter 3, the normal balance for each type of account is wherever increases are recorded.

- The normal balance for assets, expenses, and owner withdrawals are debits; therefore, the credit entries are subtracted from the debit entries to calculate the balance. If the balance is a positive number, it goes in the debit column on the trial balance. If the result is a negative number, it goes in the credit column on the trial balance.
- The normal balance for liabilities, equity, and revenue accounts are credits; therefore, the debits are subtracted from the credits to calculate the balance. If the balance is a positive number, it goes in the credit column of the trial balance. If the balance is a negative number, it goes in the debit column of the trial balance.

(Not many accounts will ever have a negative number.) Table 4-17 shows the format for a trial balance.

TABLE 4-17 ■ *Trial balance format.*

TRIAL BALANCE (NAME OF FARM OPERATION) (DATE)		
Accounts	**Debits**	**Credits**
(Assets listed first)	XXXX	
(Liabilities listed next)		XXXX
(Equity accounts listed next)		XXXX
(Revenue accounts listed next)		XXXX
(Expense accounts listed next)	XXXX	
Totals		

The last step is to add all of the numbers in the debit column and then add all of the numbers in the credit column. If the two totals are equal to each other, the trial balance is complete. If they are not equal, the farm accountant must examine all of the journal entries to ensure that they have been recorded and posted properly and that the balances have been calculated correctly.

> **Exercise 4-5** *Prepare an adjusted trial balance from the Farmers' ledger accounts in Appendix F, which include the journal entries in this chapter. The balances are calculated. Just prepare the trial balance from the information given. Then ensure that the total of the debits equals the total of the credits. Answer: Check your answer with the adjusted trial balance in Appendix G.*

> PRACTICE WHAT YOU HAVE LEARNED   *Practice what you have learned by completing Problem 4-9 at the end of the chapter.*

CHAPTER SUMMARY

Many farm operations use the cash-basis system and record revenue only when cash is received and record expenses only when cash is paid. The accrual-adjusted approach provides a more accurate picture of the financial performance and financial position of a farm business. The adjustments presented in this chapter provide the means to convert a cash-basis system to an accrual-adjusted system. These procedures allow the farm accountant to provide accrual-adjusted financial statements for lenders and other parties who need to make meaningful comparisons with the past performance of the farm business or with other farm operations.

The procedures for the adjustments discussed in this chapter include those involving inventories, prepaid expenses, depreciation expense, accrued expenses, accrued revenues, changes in values of raised breeding livestock, income taxes, and deferred taxes for the first year of operation for a farm business. After these entries are recorded and posted and an adjusted trial balance is prepared, the financial statements can be prepared, which is the subject of the next chapter.

PROBLEMS

4-1 ■ Identify which of the phrases below describe (a) the accrual-basis system, (b) the accrual-adjusted approach, or (c) the cash-basis system.

_____Revenue is recorded only when cash is received.

_____Revenue is reported when it is earned, whether the cash has been received or not.

_____Expenses are reported when they have been incurred, not necessarily when they have been paid.

_____Expenses are recorded only when cash is paid.

_____No inventory is reported.

_____Inventory is recorded when it is purchased.

_____Inventory is recorded at the end of the year.

_____Adjustments are recorded all year long.

_____Adjustments are made only when the financial statements are being prepared.

4-2 ▪ For the following scenario: a) calculate the Change in Purchased Feed Inventory and the Change in Crop Inventory, b) prepare the journal entries for the adjustments, and c) indicate the effects on the income statement for purchased feed, raised feed, and raised crops.

Steve and Chris harvested a grain crop in July and sold part of it for $13,500. The Farmers stored the rest of the crop, valued at $5,500 at the end of the year. They had no stored crop on hand at the beginning of the year. The Farmers completed the harvest of hay for their cow herd on August 1, and the value of the hay is $4,500 at harvest time. At the end of the year, they determined that $2,300 worth of that hay is still on hand on December 31 when they are preparing financial statements. Steve and Chris had no raised feed on hand at the beginning of the year. The Farmers purchased pelleted feed for $2,000 and salt and mineral blocks for $350 on October 15. This purchase was paid for on the day it was purchased and was recorded in the Purchased Feed account. The Farmers determined that the pelleted feed on hand at the end of the year cost approximately $500 and the salt and mineral blocks on hand cost approximately $250. The Farmers had no purchased feed on hand at the beginning of the year.

4-3 ▪ For the following scenario: a) calculate the Change in Purchased Feeder Livestock Inventory and the Change in Market Livestock and Poultry Inventory, b) prepare the journal entries for the adjustments, and c) indicate the effects on the income statement for purchased feeder livestock and raised market livestock.

The Farmers purchased feeder calves for $10,500 and feeder pigs for $2,500. At the end of the year, they had all of the calves and half of the feeder pigs on hand. The Farmers also had raised feeder calves on hand with a market value of $26,000 at the end of the year. The Farmers had no market livestock on hand at the beginning of the year.

4-4 ▪ For each of the events below, select the appropriate adjusting journal entry that would be recorded at the end of the year (December 31).

The Farmers purchased a one-year insurance policy on June 1 for $1800. No prepaid expenses existed at the beginning of the year.

a. Prepaid Expenses	1,050	
Change in Prepaid Insurance		1,050
b. Prepaid Expenses	750	
Change in Prepaid Insurance		750
c. Change in Prepaid Insurance	1,050	
Prepaid Expenses		1,050
d. Change in Prepaid Insurance	750	
Prepaid Expenses		750

The Farmers spent $3,000 for the costs of growing a perennial crop and recorded these costs as expenses. The crop is not harvested during the same year that it was grown. No investment in growing crops existed in the accounting records at the beginning of the year.

a. No adjusting entry is necessary under the accrual-adjusted approach.

b. Change in Investment in Growing Crops   3,000

       Cash Investment in Growing Crops   3,000

c. Cash Investment in Growing Crops   3,000

       Change in Investment in Growing Crops   3,000

4-5 ■ For each of the multiple choice questions below, select the appropriate adjusting journal entry that would be recorded at the end of the year (December 31).

The Farmers determined that their unpaid bills at the end of the year are $690. They will not pay the bills before December 31. They had no bills at the beginning of the year.

a. Accounts Payable   690

       Change in Accounts Payable   690

b. Prepaid Expense   690

       Change in Accounts Payable   690

c. Change in Accounts Payable   690

       Accounts Payable   690

d. Prepaid Expense   690

       Accounts Payable   690

The Farmers determined that they owed $3500 in real estate and property taxes at the end of the year. They will not pay the taxes before December 31. They did not owe any taxes at the beginning of the year.

a. Taxes Payable   3,500

       Change in Taxes Payable   3,500

b. Taxes Payable   3,500

       Cash   3,500

c. Change in Accounts Payable   3,500

       Accounts Payable   3,500

d. Change in Taxes Payable   3,500

       Taxes Payable   3,500

4-6 ■ Place a check mark beside each of the following events that would require an end-of-year adjustment. Assume that the end of the fiscal year is December 31.

_____Cash was paid for a utility bill before the end of the year.

_____A bill for veterinary services was received before the end of the year and will be paid the following year.

_____Grain was sold on December 21 but the check was not received by December 31.

_____A perennial crop was harvested and sold before the end of the year.

_____All feed that was purchased during the year was used up by the end of the year.

_____No purchased market livestock were on hand at the end of the year.

_____50 head of raised feeder pigs were on hand at the end of the year.

_____Raised grain for feed was still on hand at the end of the year.

4-7 ▪ For each of the multiple choice questions below, select the appropriate adjusting journal entry that would be recorded at the end of the year (December 31).

Suppose that the Farmers estimated on December 31 that they owed income tax for $3,670. They will pay their taxes in February next year. They did not owe any income taxes at the beginning of the year.

a. Taxes Payable	3,670	
Change in Taxes Payable		3,670
b. Taxes Payable	3,670	
Cash		3,670
c. Change in Taxes Payable	3,670	
Taxes Payable		3,670

d. No entry is required under the accrual-adjusted approach.

The Farmers determine that the amount of current deferred taxes at the end of the year is $2,000. No current deferred taxes existed at the beginning of the year.

a. Change in Taxes Payable	2,000	
Current Deferred Taxes		2,000
b. Current Deferred Taxes	2,000	
Change in Taxes Payable		2,000
c. Change in Taxes Payable	2,000	
Taxes Payable		2,000

d. No entry is required under the accrual-adjusted approach.

4-8 ▪ Create ledger accounts for the items in the following adjusting journal entries and post the entries to the ledger accounts.

a. Feed Inventory Purchased for Use	570	
Change in Purchased Feed Inventories		570
b. Crop Inventory Raised for Sale	10,000	
Change in Crop Inventory		10,000
c. Feeder Livestock Inventory Purchased for Resale	9,050	
Change in Purchased Feeder Livestock Inventory		9,050
d. Prepaid Expenses	1,000	
Change in Prepaid Insurance		1,000
e. Depreciation Expense	31,000	
Accumulated Depreciation		31,000
f. Change in Accounts Payable	600	
Accounts Payable		600

g. Breeding Livestock   2,100

            Gains/Losses Due to Changes in General

                  Base Values of Breeding Livestock   2,100

h. Change in Taxes Payable   3,600

            Non-Current Deferred Taxes   3,600

4-9 ■ Using the unadjusted trial balance below (from Appendix E) and the ledger accounts from Problem 4-8, prepare an adjusted trial balance.

Unadjusted Trial Balance
Steve and Chris Farmer
December 31, 20X1

Accounts	Debits	Credits
1000 Cash	$ 203.80	
1500 Breeding Livestock	73,500.00	
1600 Machinery and Equipment	65,000.00	
1650 Office Equipment and Furniture	1,000.00	
1700 Perennial Crops	45,000.00	
1800 Land, Buildings and Improvements	470,000.00	
1910 Leased Assets	100,000.00	
2100 Taxes Payable		2,160.00
2310 Notes Payable Due within One Year		50,000.00
2400 Real Notes Payable—Non-Current		120,000.00
2600 Obligations on Leased Assets		76,018.00
3100 Retained Capital		104,540.00
3110 Owner Withdrawals	150.00	
3120 Non-Farm Income		100.00
3130 Other Capital Contributions/Gifts/Inheritances		350,000.00
4000 Cash Crop Sales		12,600.00
4100 Cash Sales of Market Livestock		50,900.00
4500 Gains (Losses) from Sale of Culled Breeding Livestock	250.00	
5000 Feeder Livestock	1,500.00	
5020 Purchased Feed	1,000.00	
6100 Wage Expense	1,200.00	
6110 Payroll Tax Expense	129.20	
6310 Truck and Machinery Hire	150.00	
6520 Herbicides, Pesticides	500.00	
6630 Livestock Supplies, Tools, and Equipment	75.00	
6700 Insurance	1,200.00	
6710 Real Estate and Personal Property Taxes	1,300.00	
8100 Interest Expense	1,200.00	
8200 Gains (Losses) on Sales of Farm Capital Assets	800.00	
9100 Income Tax Expense	2,160.00	
Totals	$766,318.00	$766,318.00

CHAPTER 5

Financial Statement Preparation and Closing Entries

I n Chapter 4, you learned how to tell the difference between the cash-basis system, the accrual-basis system, and the accrual-adjusted approach; how to prepare and post adjusting journal entries for inventory, prepaid items, depreciation, accrued expenses, accrued revenues, changes in value for raised breeding livestock, income taxes, and deferred taxes in the accrual-adjusted approach; and how to prepare an adjusted trial balance.

This chapter outlines the procedures for preparing financial statements from two perspectives. One perspective continues from Chapter 4, which illustrated the recording and posting of adjusting journal entries and concluded with an adjusted trial balance. The other perspective starts with an unadjusted trial balance, then the preparation of cash-basis financial statements, followed by adjustments made directly to the financial statements. This chapter also teaches you how to prepare market-based financial statements and how to perform the closing entries after preparing financial statements.

In this chapter, you will learn how to prepare an income statement, statement of owner equity, balance sheet, and statement of cash flows from an unadjusted trial balance and from an adjusted trial balance. You will then learn how to add adjustments to cash-basis financial statements to convert them into accrual-adjusted financial statements. This chapter also teaches you how to prepare financial statements with market values. Finally, you will learn how to record closing journal entries, why it is necessary to record them, and how to calculate adjustments after the first year of operation.

PREPARING FINANCIAL STATEMENTS

So far, you have learned how to record cash transactions in the journal as they occur, and to post them to the ledger accounts. Recall that the journal is a chronological record of transactions and the ledger is a collection of all of the accounts

Learning Objective 1 ■ To list the steps involved in preparing financial statements.

involved in the transactions. You have also learned how to record journal entries for the year-end adjustments—so that the income statement will present an accurate picture of the earnings and expenses of the farm business. The balance sheet will also report a more accurate picture of assets, liabilities, and equity after the adjustments. You have also learned the importance of preparing financial statements at least once a year, to keep the financial information timely.

You can perform year-end adjustments by recording adjusting journal entries (as you learned how to do in Chapter 4) or by adjusting cash-basis financial statements directly without recording journal entries. Figure 5-1 demonstrates the differences between the two ways of making adjustments.

As Figure 5-1 illustrates, if you choose not to record adjusting journal entries, you prepare a trial balance called an **Unadjusted Trial Balance**, which does not display any adjustments. An unadjusted trial balance would look like the one in Appendix E.

Income Statement

Learning Objective 2 ■ To prepare an income statement from a trial balance.

After preparing the trial balance, the first financial statement to prepare is the income statement. As Figure 5-1 indicates, you can prepare an accrual-adjusted income statement after recording and posting adjusting journal entries. Or, you can first prepare a cash-basis income statement and then add the adjustments. Using the format for the income statement that you learned about in Chapter 1, you can prepare the income statement with the information from the revenues, expenses, gains, and losses accounts on either the adjusted or unadjusted trial balance. Figure 5-2 demonstrates the process of preparing a cash-basis income statement from the Farmers' unadjusted trial balance in Appendix E.

> **Exercise 5-1** *Using the Farmers' adjusted trial balance in Appendix G, prepare the accrual-adjusted income statement. Answer: Check your income statement with the income statement in Appendix A.*

> PRACTICE WHAT YOU HAVE LEARNED   *Practice what you have learned and complete Problems 5-1 and 5-2 at the end of the chapter.*

Statement of Owner Equity

Learning Objective 3 ■ To prepare the statement of owner equity.

The statement of owner equity summarizes the transactions that involve the equity accounts. You prepare it using the information from the trial balance and the income statement. Figure 5-3 demonstrates the process of preparing a cash-basis statement of owner equity for the Farmers using the format discussed in Chapter 1. The Farmers use the unadjusted trial balance and the net cash income from the cash-basis income statement. The amounts for owner withdrawals, non-farm income, and gifts and inheritances are from the ledger accounts listed in the unadjusted trial balance.

> **Exercise 5-2** *Using the adjusted trial balance in Appendix G and the accrual-adjusted income statement in Appendix A, prepare the statement of owner equity. Answer: Check your statement of owner equity with the one in Appendix A.*

FIGURE 5-1 ■ *Steps in the preparation of financial statements.*

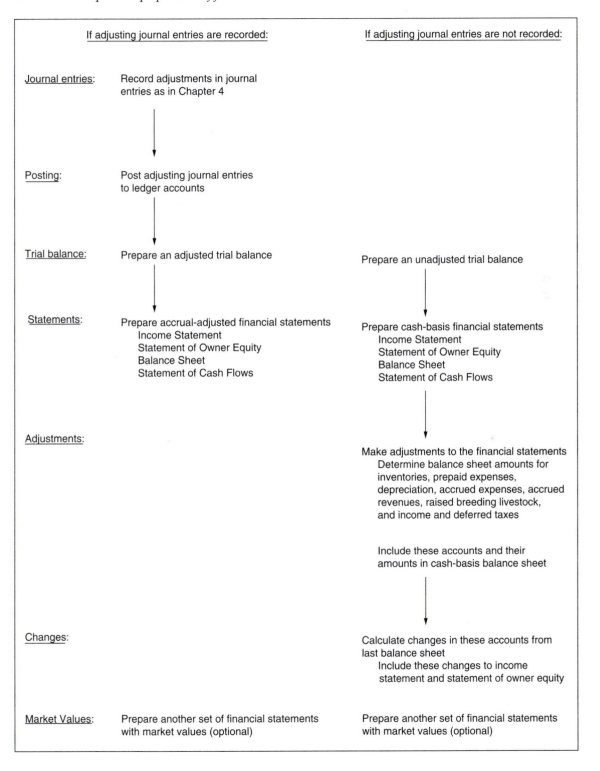

FIGURE 5-2 ■ *Preparing a cash-basis income statement from an unadjusted trial balance.*

UNADJUSTED TRIAL BALANCE Steve and Chris Farmer December 31, 20X1			INCOME STATEMENT (Cash-Basis) Steve and Chris Farmer December 31, 20X1	
Accounts	Debits	Credits		
1000 Cash	$ 203.80			
1500 Breeding Livestock	73,500.00			
1600 Machinery and Equipment	65,000.00			
1650 Office Equipment and Furniture	1,000.00			
1700 Perennial Crops	45,000.00			
1800 Land, Buildings and Improvements	470,000.00			
1910 Leased Assets	100,000.00			
2100 Taxes Payable		2,160.00		
2310 Notes Payable Due within One Year		50,000.00		
2400 Real Notes Payable—Non-Current		120,000.00		
2600 Obligations on Leased Assets		76,018.00		
3100 Retained Capital		104,540.00		
3110 Owner Withdrawals	150.00			
3120 Non-Farm Income		100.00		
3130 Other Capital Contributions/Gifts/Inheritances		350,000.00	Cash Crop Sales	$12,600.00
4000 Cash Crop Sales		12,600.00	Cash Sales of	
4100 Cash Sales of Market Livestock		50,900.00	Market Livestock	50,900.00
4500 Gains (Losses) from Sale of			Gain on Sale of Culled	
Culled Breeding Livestock	250.00		Breeding Livestock	(250.00)
5000 Feeder Livestock	1,500.00		Total Revenue	$63,250.00
5020 Purchased Feed	1,000.00		Feeder Livestock	(1,500.00)
6100 Wage Expense	1,200.00		Purchased Feed	(1,000.00)
6110 Payroll Tax Expense	129.20		Wage Expense	(1,200.00)
6310 Truck and Machinery Hire	150.00		Payroll Tax Expense	(129.20)
6520 Herbicides, Pesticides	500.00		Truck/Machine Hire	(150.00)
6630 Livestock Supplies, Tools, and Equipment	75.00		Herbicides/Pesticides	(500.00)
6700 Insurance	1,200.00		Livestock Supplies	(75.00)
6710 Real Estate and Personal Property Taxes	1,300.00		Insurance	(1,200.00)
8100 Interest Expense	1,200.00		Real Estate/Property	
8200 Gains (Losses) on Sales of Farm Capital Assets	800.00		Taxes	(1,300.00)
9100 Income Tax Expense	2,160.00		Interest Expense	(1,200.00)
			Loss on Sale	(800.00)
Totals	$766,318.00	$766,318.00	Income Tax Expense	(2,160.00)
			Net Cash Income	$52,035.80

PRACTICE WHAT YOU HAVE LEARNED *Practice what you have learned and complete Problem 5-3 at the end of the chapter.*

Balance Sheet

Learning Objective 4 ■ To prepare a balance sheet.

After preparing the statement of owner equity, the next statement to prepare is the balance sheet. Following the format for the balance sheet that you learned about in Chapter 1, you can prepare the balance sheet with the information for the asset, liability, and equity accounts on the trial balance and the owners' equity at the end of the year from the statement of owner equity. Remember from Chapter 1 that the balance sheet lists the current assets and current liabilities first in their respective sections. Figure 5-4 demonstrates the process of preparing a cash-basis balance sheet for the Farmers. We use the unadjusted trial balance and the owner equity at the end of the year from the statement of owner equity in Figure 5-3 to prepare the

FIGURE 5-3 ■ *Preparing the statement of owner equity from the trial balance and income statement.*

UNADJUSTED TRIAL BALANCE
Steve and Chris Farmer
December 31, 20X1

Accounts	Debits	Credits
1000 Cash	$ 203.80	
1500 Breeding Livestock	73,500.00	
1600 Machinery and Equipment	65,000.00	
1650 Office Equipment and Furniture	1,000.00	
1700 Perennial Crops	45,000.00	
1800 Land, Buildings and Improvements	470,000.00	
1910 Leased Assets	100,000.00	
2100 Taxes Payable		2,160.00
2310 Notes Payable Due within One Year		50,000.00
2400 Real Notes Payable—Non-Current		120,000.00
2600 Obligations on Leased Assets		76,018.00
3100 Retained Capital		104,540.00
3110 Owner Withdrawals	150.00	
3120 Non-Farm Income		100.00
3130 Other Capital Contributions/Gifts/Inheritances		350,000.00
4000 Cash Crop Sales		12,600.00
4100 Cash Sales of Market Livestock		50,900.00
4500 Gains (Losses) from Sale of Culled Breeding Livestock	250.00	
5000 Feeder Livestock	1,500.00	
5020 Purchased Feed	1,000.00	
6100 Wage Expense	1,200.00	
1110 Payroll Tax Expense	129.20	
6310 Truck and Machinery Hire	150.00	
6520 Herbicides, Pesticides	500.00	
6630 Livestock Supplies, Tools, and Equipment	75.00	
6700 Insurance	1,200.00	
6710 Real Estate and Personal Property Taxes	1,300.00	
8100 Interest Expense	1,200.00	
8200 Gains (Losses) on Sales of Farm Capital Assets	800.00	
9100 Income Tax Expense	2,160.00	
Totals	$766,318.00	$766,318.00

INCOME STATEMENT (Cash-Basis)
Steve and Chris Farmer
December 31, 20X1

Cash Crop Sales	$12,600.00	
Cash Sales of Market Livestock	50,900.00	
Gain on Sale of Culled Breeding Livestock	(250.00)	
Total Revenue	$63,250.00	
Feeder Livestock	(1,500.00)	
Purchased Feed	(1,000.00)	
Wage Expense	(1,200.00)	
Payroll Tax Expense	(129.20)	
Truck/Machine Hire	(150.00)	
Herbicides/Pesticides	(500.00)	
Livestock Supplies	(75.00)	
Insurance	(1,200.00)	
Real Estate/Property Taxes	(1,300.00)	
Interest Expense	(1,200.00)	
Loss on Sale	(800.00)	
Income Tax Expense	(2,160.00)	
Net Cash Income	$52,035.80	

STATEMENT OF OWNER EQUITY (Cash-Basis)
Steve and Chris Farmer
December 31, 20X1

Owners' Equity, Beginning		$104,540.00
Net Cash Income	$52,035.80	
Owner Withdrawals	(150.00)	
Non-Farm Income	100.00	
Other Capital Contributions/ Gifts/Inheritances	350,000.00	
Addition to Retained Capital		401,985.80
Owner Equity, End of Year		$506,525.80

FIGURE 5-4 ■ *Preparing the balance sheet from the trial balance and statement of owner equity.*

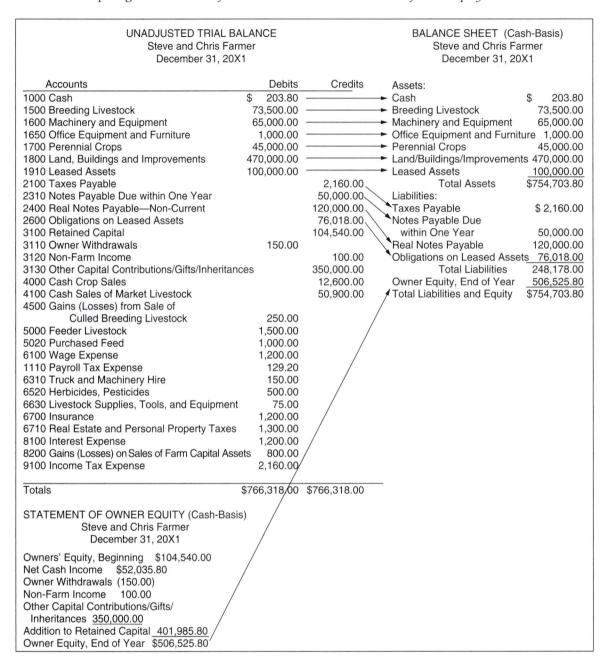

balance sheet. The amounts for the assets and liabilities are from the ledger accounts listed in the unadjusted trial balance.

Notice that the amount of Total Assets is equal to the amount of Total Liabilities and Owners' Equity; thus, the accounting equation is in balance.

Exercise 5-3 *Using the adjusted trial balance in Appendix G and the statement of owner equity in Appendix A, prepare the accrual-adjusted balance sheet. Answer: Check your balance sheet with the one in Appendix A.*

> PRACTICE WHAT YOU HAVE LEARNED   *Practice what you have learned and complete Problem 5-4 at the end of the chapter.*

Statement of Cash Flows

The statement of cash flows presents an analysis of the cash flows of the farm operation. By reporting the cash involved in financing, investing, and operating activities, the statement answers a basic question of any business operation, "Where did all the money come from and where did it go?" In particular, the statement provides valuable information about operating cash flows. Having adequate cash flows to meet operating expenses is essential to any business organization. If operating cash flows are negative, then the business is operating from either financing activities or investing activities or both. The farm business must generate positive operating cash flows to remain in business. An accrual-adjusted statement of income does not provide enough information to assess the cash position of the farm business because the "change" items are noncash in nature.

Learning Objective 5 ▄ To prepare the statement of cash flows.

To prepare the statement of cash flows, first categorize the activities that involve cash. Refer to the Cash account to categorize each transaction, beginning with operating activities, followed by investing activities, and then financing activities. Figure 5-5 displays the statement of cash flows for the Farmers.

Because this was the Farmers' first year of operation, no cash balance for the farm operation existed at the beginning of the year. Notice also that the cash balance at the end of the year is equal to the balance in the Cash account on the balance sheet. If they are not equal, you have made a mistake in preparing the statement. You must rectify the errors until the ending balance on the statement of cash flows equals the balance in the cash account.

FIGURE 5-5 ▪ *The Farmers' statement of cash flows.*

Steve and Chris Farmer	
Statement of Farm Business Cash Flows	
For the Period Ending December 31, 20X1	
Cash received from sale of livestock (other than culled breeding livestock)	$50,900.00
Cash received from sale of crops	12,600.00
Cash paid for feeder livestock	(1,500.00)
Cash paid for all other operating expenses	(3,943.00)
Cash paid for interest	(1,200.00)
Cash paid for taxes	(3,771.20)
Net cash provided by operating activities	53,085.80
Cash received from the sale of breeding livestock	2,150.00
Cash paid for purchase of machinery and equipment	(69,982.00)
Cash paid for purchase of land and buildings and improvements	(120,000.00)
Cash paid for purchase of investments	(45,000.00)
Net cash used by investing activities	(232,832.00)
Proceeds from real estate and other term loans	170,000.00
Cash received from contributions by owners	20,100.00
Principal payments for loans	(10,000.00)
Owner withdrawals	(150.00)
Net cash provided by financing activities	179,950.00
Net increase in cash from operating, investing, and financing activities	$ 203.80
Cash balance at beginning of year	0
Cash balance at end of year	$ 203.80

Exercise 5-4 *Analyze the Farmers' cash account in Appendix E and verify how they determined the dollar amounts for each of the activities in the statement of cash flows in Figure 5-5. Answer: Your first step is to classify each of the transactions.*

	1000 CASH			
Date	Description	JE#	Debits	Credits
Jan. 2	Set up farm bank account	(1)	20,000	Financing
Feb. 1	Income tax payment	(31)		2,160	Operating
Feb. 12	Transfer from personal checking account	(3)	100	Financing
Mar. 1	Money borrowed to purchase land	(6)	120,000	Financing
Mar. 1	Money borrowed to purchase tractor	(7)	50,000	Financing
Mar. 1	Purchase of land	(9)		120,000	Investing
Mar. 1	Purchase of tractor	(11)		46,000	Investing
Mar. 10	Sale of bull to neighbor	(10)	900	Investing
Mar. 17	Transfer to personal checking account	(4)		150	Financing
Apr. 1	Employee paycheck	(26)		1,018	Operating
Apr. 10	FICA taxes and FIT paid to IRS	(27)		274	Operating
Apr. 10	FUTA tax paid to IRS	(28)		9.60	Operating
Apr. 10	SUTA tax paid to state agency	(29)		27.60	Operating
Apr. 17	Purchase of cattle tags	(22)		75	Operating
May 10	Purchase of feed	(20)		1,000	Operating
May 16	Purchase of feeder pigs	(15)		1,500	Operating
May 20	Investment in apple orchard	(14)		45,000	Investing
June 1	Payment for herbicide	(23)		500	Operating
July 1	Payment of real estate and property taxes	(30)		1,300	Operating
July 17	Sale of half of the feeder pigs	(16)	900	Operating
Aug. 1	Payment of insurance premium	(25)		1,200	Operating
Aug. 1	Lease payment on harvester	(13)		23,982	Investing
Aug. 16	Sale of grain at harvest time	(17)	12,000	Operating
Oct. 1	To record interest and principal payment	(8)		11,200	Financing and Operating
Oct. 31	Sale of grown hay to neighbor	(21)	600	Operating
Nov. 20	Sale of culled cows	(19)	1,250	Investing
Dec. 1	Truck expense to haul feeder cattle	(24)		150	Operating
Dec. 1	Sale of feeder cattle	(18)	50,000	Operating

Your next step is to analyze each type of activity and identify the inflows and outflows within each activity.

Feb. 1	Income tax payment	(31)	(2,160)	Operating—Cash paid for taxes
Apr. 1	Employee paycheck	(26)	(1,018)	Operating—Cash paid for operating expenses
Apr. 10	FICA taxes and FIT paid to IRS	(27)	(274)	Operating—Cash paid for taxes
Apr. 10	FUTA tax paid to IRS	(28)	(9.60)	Operating—Cash paid for taxes
Apr. 10	SUTA tax paid to state agency	(29)	(27.60)	Operating—Cash paid for taxes
Apr. 17	Purchase of cattle tags	(22)	(75)	Operating—Cash paid for operating expenses
May 10	Purchase of feed	(20)	(1,000)	Operating—Cash paid for operating expenses

ADJUS

If you d

basis sta

amount

deprecia

Suppo

in Cha

as sho

record

items

1. Inve

   on

   han

2. Pre

   the

3. Acc

   in

4. Acc

5. Aft

   acc

   are

They

For the

also d

year is

the ye

Accou

Feed

Feed

Crop

Feede

   Pur

Prepa

Cash

Accou

Taxes

Intere

They

Yo

increas

include

Date	Description		Amount	Category
May 16	Purchase of feeder pigs	(15)	(1,500)	Operating—Cash paid for feeder livestock
June 1	Payment for herbicide	(23)	(500)	Operating—Cash paid for operating expenses
July 1	Payment of real estate and property taxes	(30)	(1,300)	Operating—Cash paid for taxes
July 17	Sale of half of the feeder pigs	(16)	900	Operating—Cash received (sale of livestock)
Aug. 1	Payment of insurance premium	(25)	(1,200)	Operating—Cash paid for operating expenses
Aug. 16	Sale of grain at harvest time	(17)	12,000	Operating—Cash received (sale of crops)
Oct. 1	To record interest and principal payment	(8)	(1,200)	Operating—Cash paid for interest
Oct. 31	Sale of grown hay to neighbor	(21)	600	Operating—Cash received (sale of crops)
Dec. 1	Truck expense to haul feeder cattle	(24)	(150)	Operating—Cash paid for operating expenses
Dec. 1	Sale of feeder cattle	(18)	50,000	Operating—Cash received (sale of livestock)
Mar. 1	Purchase of land	(9)	(120,000)	Investing—Cash paid (land/buildings)
Mar. 1	Purchase of tractor	(11)	(46,000)	Investing—Cash paid (machinery/equipment)
Mar. 10	Sale of bull to neighbor	(10)	900	Investing—Cash received (breeding livestock)
May 20	Investment in apple orchard	(14)	(45,000)	Investing—Cash paid (investments)
Aug. 1	Lease payment on harvester	(13)	(23,982)	Investing—Cash paid (machinery/equipment)
Nov. 20	Sale of culled cows	(19)	1,250	Investing—Cash received (breeding livestock)
Jan. 2	Set up farm bank account	(1)	20,000	Financing—Cash received from owners
Feb. 12	Transfer from personal checking account	(3)	100	Financing—Cash received from owners
Mar. 1	Money borrowed to purchase land	(6)	120,000	Financing—Proceeds from loans
Mar. 1	Money borrowed to purchase tractor	(7)	50,000	Financing—Proceeds from loans
Mar. 17	Transfer to personal checking account	(4)	150	Financing—Owner withdrawals
Oct. 1	To record interest and principal payment	(8)	10,000	Financing—Principal payments for loans

Then, combine the transactions that are alike.

Cash received from sale of livestock:

900	Operating—Cash received (sale of livestock)
50,000	Operating—Cash received (sale of livestock)
$50,900	

Cash received from sale of crops:

12,000	Operating—Cash received (sale of crops)
600	Operating—Cash received (sale of crops)
$12,600	

Cash paid for feeder livestock:

(1,500)	Operating—Cash paid for feeder livestock
($1,500)	

Cash paid for all other operating expenses:

(1,018)	Operating—Cash paid for operating expenses
(75)	Operating—Cash paid for operating expenses
(1,000)	Operating—Cash paid for operating expenses
(500)	Operating—Cash paid for operating expenses
(1,200)	Operating—Cash paid for operating expenses
(150)	Operating—Cash paid for operating expenses
($3,943)	

The Farmers begin the closing process by examining the trial balance to identify the accounts that need to be closed. If they did not record the adjusting entries as illustrated in Chapter 4, they would look to the unadjusted trial balance for the income statement and equity accounts that should be closed. The accounts in boldface are the income statement accounts that need to be closed.

UNADJUSTED TRIAL BALANCE
Steve and Chris Farmer
December 31, 20X1

Accounts	Debits	Credits
1000 Cash	$ 203.80	
1500 Breeding Livestock	73,500.00	
1600 Machinery and Equipment	65,000.00	
1650 Office Equipment and Furniture	1,000.00	
1700 Perennial Crops	45,000.00	
1800 Land, Buildings and Improvements	470,000.00	
1910 Leased Assets	100,000.00	
2100 Taxes Payable		2,160.00
2310 Notes Payable Due within One Year		50,000.00
2400 Real Notes Payable—Non-Current		120,000.00
2600 Obligations on Leased Assets		76,018.00
3100 Retained Capital		104,540.00
3110 Owner Withdrawals	150.00	
3120 Non-Farm Income		100.00
3130 Other Capital Contributions/Gifts/Inheritances		350,000.00
4000 Cash Crop Sales		**12,600.00**
4100 Cash Sales of Market Livestock		**50,900.00**
4500 Gains (Losses) from Sale of Culled Breeding Livestock	**250.00**	
5000 Feeder Livestock	**1,500.00**	
5020 Purchased Feed	**1,000.00**	
6100 Wage Expense	**1,200.00**	
6110 Payroll Tax Expense	**129.20**	
6310 Truck and Machinery Hire	**150.00**	
6520 Herbicides, Pesticides	**500.00**	
6630 Livestock Supplies, Tools, and Equipment	**75.00**	
6700 Insurance	**1,200.00**	
6710 Real Estate and Personal Property Taxes	**1,300.00**	
8100 Interest Expense	**1,200.00**	
8200 Gains (Losses) on Sales of Farm Capital Assets	**800.00**	
9100 Income Tax Expense	**2,160.00**	
Totals	$766,318.00	$766,318.00

The Farmers can see from the trial balance that the accounts with credit balances that need to be closed are the sales accounts. They debit these accounts in the closing journal entry. The remaining accounts that need to be closed are the loss and expense accounts with debit balances. They credit these accounts in the closing journal entry. The following closing entry would be prepared to close these accounts.

(47)	Dec. 31	4000 Cash Crop Sales	12,600	
		4100 Cash Sales of Market Livestock	50,900	
		4500 Loss from Sale of Culled Breeding Livestock		250
		5000 Feeder Livestock		1,500
		5020 Purchased Feed		1,000
		6100 Wage Expense		1,200
		6110 Payroll Tax Expense		129.20
		6310 Truck and Machinery Hire		150
		6520 Herbicides, Pesticides		500
		6630 Livestock, Tools and Equipment		75
		6700 Insurance		1,200
		6710 Real Estate and Personal Property Taxes		1,300
		8100 Interest Expense		1,200
		8200 Loss on Sales of Farm Capital Assets		800
		9100 Income Tax Expense		2,160
		3100 Retained Capital		52,035.80

The credit to Retained Capital is the amount needed to make sure that the debits equal the credits. This amount is also the Net Cash Income shown on the cash-basis Income Statement. This closing entry adds Net Cash Income to Retained Capital.

The Farmers also close the equity accounts to Retained Capital by debiting those accounts with credit balances and crediting those accounts with debit balances.

(48)	Dec. 31	3120 Non-Farm Income	100	
		3130 Other Capital Contributions		
		Gifts/Inheritances	350,000	
		3110 Owner Withdrawals		150
		3100 Retained Capital		349,950

This closing entry adds the non-farm income and capital contributions to and subtracts owner withdrawals from Retained Capital. Notice that the balance in the Retained Capital account ($506,525.80) corresponds with the amount shown on the cash-basis statement of owner equity for Retained Capital.

3100 Retained Capital					
Date	Description	JE#	Debits	Credits	Balance
Beginning Balance					0
Jan. 2	To set up farm equity	(1)		116,700	116,700
Jan. 2	To set up farm equity	(2)	12,160		104,540
Dec. 31	To close income statement accounts	(47)		52,035.80	156,575.80
Dec. 31	To close equity accounts	(48)		349,950	506,525.80

If you choose to record adjusting journal entries to make year-end adjustments, you need to include all of the income statement accounts in the closing entries, including the "change" items and other accounts used for adjustments. You would close all accounts beginning with 4, 5, 6, 7, 8, and 9 as illustrated above, including the "change" accounts. These accounts should not carry over any information from the previous year and should begin the next year with a zero balance.

Exercise 5-5   *Using the adjusted trial balance for the Farmers, prepare the closing journal entries for the income statement accounts and the equity accounts that need to be closed. Verify that the credit to Retained Earnings in the first closing entry is the same amount as Accrual-Adjusted Net Income on the accrual-adjusted income statement. Verify that the balance in the Retained Capital account after both closing entries are posted is the same as the amount on the accrual-adjusted statement of owner equity. Answer:*

ADJUSTED TRIAL BALANCE
Steve and Chris Farmer
December 31, 20X1

Accounts	Debits	Credits
1000 Cash	$203.80	
1100 Accounts Receivable	3,500.00	
1212 Feeder Livestock Inventory Purchased for Resale	750.00	
1223 Feed Inventory Raised for Use	2,300.00	
1224 Feed Inventory Purchased for Use	230.00	
1231 Crop Inventory Raised for Sale	6,000.00	
1300 Prepaid Expenses	700.00	
1400 Cash Investment in Growing Crop	2,500.00	
1500 Breeding Livestock	79,500.00	
1600 Machinery and Equipment	65,000.00	
1650 Office Equipment and Furniture	1,000.00	
1700 Perennial Crops	45,000.00	
1800 Land, Buildings and Improvements	470,000.00	
1910 Leased Assets	100,000.00	
1980 Accumulated Depreciation		22,150.00
2000 Accounts Payable		340.00
2100 Taxes Payable		3,030.00
2200 Interest Payable		5,000.00
2310 Notes Payable Due within One Year		50,000.00
2400 Real Notes Payable—Non-Current		120,000.00
2500 Non-Current Deferred Taxes		3,300.00
2510 Current Deferred Taxes		1,450.00
2600 Obligations on Leased Assets		76,018.00
3100 Retained Capital		104,540.00
3110 Owner Withdrawals	150.00	
3120 Non-Farm Income		100.00
3130 Other Capital Contributions/Gifts/Inheritances		350,000.00
4000 Cash Crop Sales		12,600.00
4010 Changes in Crop Inventories		8,300.00

4100 Cash Sales of Market Livestock		50,900.00
4500 Gains (Losses) from Sale of Culled Breeding Livestock	250.00	
4600 Change in Value Due to Change in Quantity of Raised Breeding Livestock		6,000.00
4700 Change in Accounts Receivable		3,500.00
5000 Feeder Livestock	1,500.00	
5010 Changes in Purchased Feeder Livestock Inventory		750.00
5020 Purchased Feed	1,000.00	
5030 Changes in Purchased Feed Inventory		230.00
6100 Wage Expense	1,200.00	
6110 Payroll Tax Expense	129.20	
6310 Truck and Machinery Hire	150.00	
6520 Herbicides, Pesticides	500.00	
6630 Livestock Supplies, Tools, and Equipment	75.00	
6700 Insurance	1,200.00	
6710 Real Estate and Personal Property Taxes	1,300.00	
6780 Depreciation Expense	22,150.00	
6810 Change in Accounts Payable	340.00	
6820 Change in Prepaid Insurance		700.00
6830 Change in Investment in Growing Crop		2,500.00
8100 Interest Expense	1,200.00	
8110 Change in Interest Payable	5,000.00	
8200 Gains (Losses) on Sales of Farm Capital Assets	800.00	
9100 Income Tax Expense	2,160.00	
9110 Change in Taxes Payable	5,620.00	
Totals	$821,408.00	$821,408.00

First closing entry:

(47)	Dec. 31	4000 Cash Crop Sales	12,600	
		4010 Change in Crop Inventory	8,300	
		4100 Cash Sales of Market Livestock	50,900	
		4600 Change in Value Due to Change in		
		Quantity of Raised Breeding Livestock	6,000	
		4700 Change in Accounts Receivable	3,500	
		5010 Changes in Purchased Feeder Livestock Inventory	750	
		5030 Changes in Purchased Feed Inventory	230	
		6820 Change in Prepaid Insurance	700	
		6830 Change in Investment in Growing Crop	2,500	
		4500 Loss from Sale of Culled Breeding Livestock		250
		5000 Feeder Livestock		1,500
		5020 Purchased Feed		1,000
		6100 Wage Expense		1,200
		6110 Payroll Tax Expense		129.20
		6310 Truck and Machinery Hire		150
		6520 Herbicides, Pesticides		500
		6630 Livestock, Tools and Equipment		75
		6700 Insurance		1,200
		6710 Real Estate and Personal Property Taxes		1,300

6780 Depreciation Expense	22,150
6810 Change in Accounts Payable	340
8100 Interest Expense	1,200
8110 Change in Interest Payable	5,000
8200 Loss on Sales of Farm Capital Assets	800
9100 Income Tax Expense	2,160
9110 Change in Taxes Payable	5,620
3100 Retained Capital	40,905.80

Verify that the credit to Retained Earnings equals Accrual-Adjusted Net Income:

ACCRUAL-ADJUSTED INCOME STATEMENT

Steve and Chris Farmer

December 31, 20X1

Cash Crop Sales	$12,600.00
Change in Crop Inventories	8,300.00
Cash Sales of Market Livestock	50,900.00
Loss from Sale of Culled Breeding Livestock	(250.00)
Change in Value Due to Change in Quantity of Raised Breeding Livestock	6,000.00
Change in Accounts Receivable	3,500.00
Gross Revenues	$81,050.00
Feeder Livestock	(1,500.00)
Changes in Purchased Feeder Livestock Inventory	750.00
Purchased Feed	(1,000.00)
Changes in Purchased Feed Inventory	230.00
Wage Expense	(1,200.00)
Payroll Tax Expense	(129.20)
Truck and Machinery Hire	(150.00)
Herbicides, Pesticides	(500.00)
Livestock Supplies	(75.00)
Insurance	(1,200.00)
Real Estate and Personal Property Taxes	(1,300.00)
Depreciation Expense	(22,150.00)
Change in Accounts Payable	(340.00)
Change in Prepaid Insurance	700.00
Change in Investment in Growing Crop	2,500.00
Interest Expense	(1,200.00)
Change in Interest Payable	(5,000.00)
Net Farm Income from Operations	$49,485.80
Loss on Sales of Farm Capital Assets	(800.00)
Income before Taxes	$48,685.80
Income Tax Expense	(2,160.00)
Change in Taxes Payable	(870.00)
Change in Taxes Payable	(1,450.00)
Change in Taxes Payable	(3,300.00)
Accrual Adjusted Net Income	$40,905.80

Second closing entry:

(48)	Dec. 31	3120 Non-Farm Income	100	
		3130 Other Capital Contributions/Gifts/Inheritances	350,000	
		3110 Owner Withdrawals		150
		3100 Retained Capital		349,950

Verify that the balance in the Retained Capital account equals Owner Equity at the end of the year on Statement of Owner Equity:

3100 RETAINED CAPITAL					
Date	**Description**	**JE#**	**Debits**	**Credits**	**Balance**
Beginning Balance					0
Jan. 2	To set up farm equity	(1)		116,700	116,700
Jan. 2	To set up farm equity	(2)	12,160		104,540
Dec. 31	To close income statement accounts	(47)		40,905.80	145,445.80
Dec. 31	To close equity accounts	(48)		349,950	495,395.80

ACCRUAL-ADJUSTED STATEMENT OF OWNER EQUITY
Steve and Chris Farmer
December 31, 20X1

Owners' Equity, Beginning	$104,540.00
Accrual-Adjusted Net Income	40,905.80
Owner Withdrawals	(150.00)
Non-Farm Income	100.00
Other Capital Contributions/Gifts/Inheritances	350,000.00
Owner Equity, End of Year	$495,395.80

Because of closing entries (47) and (48), the accounts now have a zero balance. Appendix H displays the income statement and equity accounts after posting the closing entries.

PRACTICE WHAT YOU HAVE LEARNED	*Practice what you have learned and complete Problem 5-8*
at the end of the chapter.	

ADJUSTMENTS FOR SUBSEQUENT YEARS

Adjustments in subsequent years for accrual-adjusted financial statements must take into account the amounts of inventories, prepaid expenses, accrued expenses, accrued revenues, income taxes, and deferred taxes from the previous year and adjust for the differences in these items from the previous year to the current year. The following examples demonstrate the procedures for recording adjusting journal entries for the Farmers' second year of operation. If directly adjusting the cash-basis financial statements, the computations are the same as those in these examples.

Learning Objective 9 ■ To determine the amount of the adjustments in subsequent years

Inventories

The following transactions and events occurred for the Farmers in the year 20X2:

- The Farmers purchased $1,000 worth of feed in 20X2. The amount remaining at the end of the year had a cost and market value of $100.
- Steve and Chris harvested hay in 20X2, and $2,500 worth of that hay is still on hand on December 31, 20X2.
- The value of raised crops at the end of 20X2 is $5,000.
- They sold the remaining feeder pigs at the beginning of 20X2 and had no feeder pigs on hand at the end of the year.
- During 20X2, the Farmers raised a new calf crop. The Farmers sold these calves in 20X2 before the end of the year for $47,000 and did not have any feeder cattle on hand at the end of the year.

For the purchased feed, the Change in Purchased Feed Inventory would be the decrease between the current value of $100 and the cost of $230 reported in 20X1. The market-based income statement reports the difference between the current value of $100 and last year's market value of $200.

	Cost	Market
Amount of purchased feed left over	$ 100	$ 100
– Feed Inventory Purchased for Use at beginning of the year	(230)	(200)
= Change in Purchased Feed Inventory	$(130)	$ (100)

If you make adjustments by recording journal entries, you record the following adjustment for the cost values:

(32a)   Dec. 31   5030 Change in Purchased Feed Inventories   130
                            1224 Feed Inventory Purchased for Use   130

1224 FEED INVENTORY PURCHASED FOR USE						
Date	Description	JE#	Debits	Credits	Balance
Beginning Balance					0
Dec. 31	Adjusting entry for value of purchased feed on hand	(32)	230		230
Dec. 31	Adjusting entry for value of purchased feed on hand	(32a)		130	100

		Cost	Market
Income Statement:			
	Gross Revenue	$XXX	$XXX
	– Purchased Feed	(1,000)	(1,000)
	+ Change in Purchased Feed Inventory	(130)	(100)
	Net Effect on Net Farm Income from Operations	($1,130)	($1,100)

Balance Sheet:

	Cost	Market
Assets:		
Feed Inventory Purchased for Use	$100	$100

This adjusting entry adjusts the balance of Feed Inventory Purchased for Use down to $100.

For the raised feed, the Change in Crop Inventory would be the increase between the current value of $2,500 and the value of $2,300 reported in 20X1. Because these values for raised feed are based on market values, the market-based financial statements will report the same results as the cost-based financial statements.

	Cost	Market
Amount of raised feed left over at the end of the year	$2,500	$2,500
– Feed Inventory Raised for Use at beginning of the year	(2,300)	(2,300)
= Change in Crop Inventory	$ 200	$ 200

Using journal entries, you would adjust the amount of raised feed on hand in the following way.

(33a)   Dec. 31   1223 Feed Inventory Raised for Use   200
                        4010 Change in Crop Inventory   200

1223 FEED INVENTORY RAISED FOR USE					
Date	Description	JE#	Debits	Credits	Balance
Beginning Balance					0
Dec. 31	Adjusting entry for market value of hay on hand	(33)	2,300		2,300
Dec. 31	Adjusting entry for market value of hay on hand	(33a)	200		2,500

Income Statement:	Cost	Market
Cash Crop Sales	$XXX	$XXX
+ Change in Crop Inventory ($2,500 – 2,300)	200	200
Effect on Gross Revenue	$ 200	$ 200

Balance Sheet:		
Assets:		
Feed Inventory Raised for Use	$2,500	$2,500

This adjusting entry adjusts the balance of Feed Inventory Raised for Use up to $2,500.

You would perform similar procedures at the end of 20X2 for crops raised for sale. In 20X1, Steve and Chris stored part of their grain crop, valued at $6,000 (journal entry (34)). In 20X2, they sold last year's crop for its market value of $6,000 and this year's crop has a value of only $5,000. They make an adjustment on the income statement as Change in Crop Inventory for the decrease in value from $6,000 to $5,000.

	Cost	Market
Amount of raised crop left over at the end of the year	$ 5,000	$ 5,000
– Crop Inventory Raised for Sale at beginning of the year	(6,000)	(6,000)
= Change in Crop Inventory	$(1,000)	$(1,000)

Journal entries adjust the amount of the crop on hand in the following way.

(34a)   Dec. 31   4010 Change in Crop Inventory   1,000
                        1231 Crop Inventory Raised for Sale   1,000

1231 CROP INVENTORY RAISED FOR SALE					
Date	**Description**	**JE#**	**Debits**	**Credits**	**Balance**
Beginning Balance					0
Dec. 31	Adjusting entry for market value of crop on hand	(34)	6,000		6,000
Dec. 31	Adjusting entry for market value of crop on hand	(34a)		1,000	5,000

Income Statement:

	Cost	Market
Cash Crop Sales	$6,000	$6,000
+ Change in Crop Inventory ($5,000 – 6,000)	(1,000)	(1,000)
Effect on Gross Revenue	$5,000	$5,000

Balance Sheet:
    Assets:

	Cost	Market
Crop Inventory Raised for Sale	$5,000	$5,000

For the feeder pigs purchased for resale, the adjustment is for the difference between the current value of $0 and the $750 cost of the purchased pigs that were on hand at the end of 20X1. Market-based financial statements report the difference between the current value of $0 and last year's market value of $900.

	Cost	Market
Amount of purchased market livestock left over at the end of the year	$ 0	$ 0
– Feeder Livestock Purchased for Resale at beginning of the year	(750)	(700)
= Change in Purchased Feeder Livestock Inventory	$(750)	$(700)

Using journal entries, you would record the adjustment as follows just before you prepare the financial statements.

(35a) Dec. 31   5010 Change in Purchased Feeder   750
                    Livestock Inventory
                      1212 Feeder Livestock Inventory Purchased for Resale   750

1212 FEEDER LIVESTOCK INVENTORY PURCHASED FOR RESALE					
Date	**Description**	**JE#**	**Debits**	**Credits**	**Balance**
Beginning Balance					0
Dec. 31	Adjusting entry for value of purchased feeder pigs	(35)	750		750
Dec. 31	Adjusting entry for value of purchased feeder pigs	(35a)		750	0

Income Statement:

	Cost	Market
Operating Expenses:		
Feeder Livestock	$ 0	$ 0
+ Change in Purchased Feeder Livestock Inventory	(750)	(700)
Net Effect on Net Farm Income from Operations	$(750)	$(700)

Because of these adjustments, Feeder Livestock Inventory Purchased for Resale has a $0 balance and the balance sheet does not show it.

In the year 20X1, the Farmers raised and sold feeder calves. Therefore, they did not report any feeder livestock raised for sale on the balance sheet that year. In 20X2, they raised and sold another calf crop and again did not have any raised feeder livestock on hand at the end of the year. No adjustment is necessary because they have no inventory to report. The income statement merely reports the sales that occurred.

Prepaid Expenses

The difference in prepaid expenses at the end of last year and the end of this year is the amount of the adjustments made in subsequent years.

- The amount of insurance premiums paid by the Farmers in 20X2 is the same as they were in 20X1.
- They sold the 20X1 crop in 20X2. They harvested a new apple crop in 20X2 and did not sell it by the end of the year. The new apple crop incurred costs of $2,750 in 20X2.

In the year 20X1, Steve and Chris purchased insurance for $1,200 for a one-year policy on August 1 (journal entry (25)). Next year on August 1, they record another insurance premium payment in the same way and, if there are no changes to the amount of the premium, the amount of prepaid insurance at the end of the year will be the same as the previous year. The Prepaid Insurance account balance will be $700 at the end of 20X2 and at the end of each following year until the amount of the premium changes or they drop the policy. Because there is no difference in prepaid expenses, no adjustment for Change in Prepaid Insurance is required. The income statement reports the amount of cash paid for insurance premiums, which is the annual amount of insurance expense.

1300 PREPAID EXPENSES					
Date	Description	JE#	Debits	Credits	Balance
Beginning Balance					0
Dec. 31	Adjusting entry for insurance paid in advance	(37)	700		700

Income Statement: Cost and Market

Operating Expenses:

Insurance	$(1,200)
+ Change in Prepaid Insurance	0
Net Effect on Net Farm Income, Accrual Adjusted	$(1,200)

For the apple crop from the orchard, the income statement must report the cash sale of last year's crop, the cash expenditures for this year's crop ($2,750), and the difference in Cash Investment in Growing Crops between 20X1 and 20X2 (an increase of $250). If the cost of the investment is the same as the market value, the market-based financial statement will report the same numbers as the cost-based financial statements. If last year's crop was sold for $2,900, they would report the following effects:

	Cost	Market
Amount of the Cash Investment in Growing Crops at the end of the year	$2,750	$2,750
Cash Investment in Growing Crops at the beginning of the year	(2,500)	(2,500)
= Change in Investment in Growing Crops	$ 250	$ 250

Using journal entries, they record the following adjusting entry before financial statements are prepared to report the Change in Investment in Growing Crops.

(38a) Dec. 31   1400 Cash Investment in Growing Crops   250
                 6830 Change in Investment in Growing Crops   250

1400 CASH INVESTMENT IN GROWING CROPS					
Date	**Description**	**JE#**	**Debits**	**Credits**	**Balance**
Beginning Balance					0
Dec. 31	Adjusting entry for expenditures in orchard	(38)	2,500		2,500
Dec. 31	Adjusting entry for expenditures in orchard	(38a)	250		2,750

Income Statement:

		Cost and Market
Cash Crop Sales		$2,900
Effect on Gross Revenue		$2,900
Operating Expenses		
+ Change in Investment in Growing Crops	250	
Net Effect on Operating Expenses		250
Net Effect on Net Farm Income from Operations		$3,150

Balance Sheet:
    Assets:
        Cash Investment in Growing Crops $2,750

Another approach to this situation is to realize the net effect from the sale of the crop. The revenue from the 20X1 crop is $2,900 and the expense of producing that crop was $2,500 for a profit of $400. The additional positive effect on net farm income is the investment of $2,750 for the 20X2 crop. The $400 profit plus the $2,750 investment equals $3,150, the amount reported on the income statement.

Accrued Expenses

They report the differences in accrued expenses from the previous year to the current year. The procedures are similar to the procedures for inventories and prepaid expenses.

- The Farmers owed $500 in interest at the end of 20X2.
- The Farmers owed $500 in unpaid bills at the end of 20X2.

In Chapter 3, Steve and Chris borrowed $50,000 to purchase a tractor on March 1, 20X1 (journal entry (7)). They make the payment a year later and pay off the entire note plus interest of $6,000. They report the payment for interest as Interest Expense. Suppose that Steve and Chris borrowed additional money during the year 20X2 and the amount of accrued interest on the new loan as of December 31, 20X2 was $500. An adjustment is required on the income statement for the difference between the $5,000 Interest Payable reported on the 20X1 balance sheet and the $500 interest owed at the end of 20X2.

Amount of the interest owed at the end of the year	$ 500
− Interest Payable at the beginning of the year	(5,000)
= Change in Interest Payable	$(4,500)

Using journal entries, they would record the following adjusting journal entry before preparing the 20X2 financial statements.

(40a) Dec 31   2200 Interest Payable   4,500
                    5810 Change in Interest Payable   4,500

	2200 INTEREST PAYABLE				
Date	**Description**	**JE#**	**Debits**	**Credits**	**Balance**
Beginning Balance					0
Dec. 31	Adjusting entry to record change in interest payable	(40)		5,000	5,000
Dec. 31	Adjusting entry to record change in interest payable	(40a)	4,500		500

Income Statement:

Interest Expense	$(6,000)
– Change in Interest Payable	4,500
Net Effect on Net Farm Income, Accrual Adjusted	$(1,500)

Balance Sheet:
    Liabilities:
        Interest Payable $ 500

At the end of 20X2, and at the end of every year thereafter, the Farmers need to determine the amount of unpaid bills and report the amount on the balance sheet as Accounts Payable. The amount for the Change in Accounts Payable is the difference between the amount of Accounts Payable of the previous year ($340) and the amount of Accounts Payable at the end of the current year ($500). Using journal entries, they would record the following adjusting journal entry before preparing financial statements.

(41a) Dec. 31   6810 Change in Accounts Payable   160
                      2000 Accounts Payable   160

	2000 ACCOUNTS PAYABLE				
Date	**Description**	**JE#**	**Debits**	**Credits**	**Balance**
Beginning Balance					0
Dec. 31	Adjusting entry to record change in accounts payable	(41)		340	340
Dec. 31	Adjusting entry to record change in accounts payable	(41a)		160	500

Income Statement:

Operating Expenses	
+ Change in Accounts Payable	$(160)
Net effect on Net Farm Income from Operations	$(160)

Balance Sheet:
    Liabilities:
        Accounts Payable $ 500

As indicated in Chapter 4, they would perform the preceding procedures for any unpaid taxes or other current liabilities. They compare the amount currently owed to the amount owed at the end of last year and record the difference on the income statement. The balance sheet reports the correct amount of liabilities.

Accrued Revenues

You also make adjustments for money owed to the farm business and for changes in value for raised breeding livestock.

- The Farmers were owed $5,000 at the end of 20X2.
- The increase in value of raised breeding livestock at the end of 20X2 was $3,000.

The income statement reports the difference between the Accounts Receivable of $3,500 at the end of 20X1 and the $5,000 in Accounts Receivable at the end of 20X2 as Change in Accounts Receivable.

Amount of money owed to the farm business at the end of the year	$5,000
− Accounts Receivable at beginning of the year	(3,500)
= Change in Accounts Receivable	$1,500

Using journal entries, they record the following adjusting journal entry before they prepare the financial statements.

(42a) Dec. 31 1100 Accounts Receivable   1,500
                4700 Change in Accounts Receivable   1,500

1100 ACCOUNTS RECEIVABLE					
Date	**Description**	**JE#**	**Debits**	**Credits**	**Balance**
Beginning Balance					0
Dec. 31	Adjusting entry to record change in accounts receivable	(42)	3,500		3,500
Dec. 31	Adjusting entry to record change in accounts receivable	(42a)	1,500		5,000

Income Statement:
           Cash Sales of Market Livestock and Poultry $XXX
           + Change in Accounts Receivable   1,500
              Effect on Gross Revenue $1,500

Balance Sheet:
      Assets:
      Accounts Receivable   5,000

If Steve and Chris had determined that the change in the value of their raised cattle herd due to age progression was an increase of $3,000 from 20X1 to 20X2, they would report an adjustment on the balance sheet by adding $3,000 to the base values reported at the beginning of the year. They adjust the income statement for Change in Value Due to Change in Quantity of Raised Breeding Livestock for the same amount. They record the following adjusting journal entry so that the balance sheet reports the adjusted value for the breeding livestock.

(43a) Dec. 31 1500 Breeding Livestock   3,000
                4600 Change in Value Due to Change in
                     Quantity of Raised Breeding Livestock   3,000

1500 BREEDING LIVESTOCK					
Date	**Description**	**JE#**	**Debits**	**Credits**	**Balance**
Beginning Balance					0
Jan. 2	Set up farm account for breeding cattle	(1)	76,000		76,000
Mar. 10	Sale of bull to neighbor	(10)		1,000	75,000
Nov. 20	Sale of culled cows	(19)		1,500	73,500
Dec. 31	Adjustment for change in quantity of raised breeding cows	(43)	6,000		79,500
Dec. 31	Adjustment for change in quantity of raised breeding cows	(43a)	3,000		82,500

Income Statement:

+ Change in Value Due to Change in Quantity of $3,000
  Raised Breeding Livestock
        Effect on Gross Revenue $3,000

Balance Sheet:
        Assets:
            Breeding Livestock $82,500

Each year the progression of animals from one age group to another is assessed and the appropriate adjustments are made. Each year you also need to assess whether or not changes in base value are required. Chapter 8 discusses more details on these topics and procedures.

Income Taxes and Deferred Taxes

Adjustments are required for income taxes owed and deferred taxes. The differences between these tax liabilities at the end of the previous year and the end of the current year are reported. Each year, the balance sheet reports the estimated amounts for income taxes owed and deferred taxes. Chapter 9 discusses these procedures in more detail.

> **PRACTICE WHAT YOU HAVE LEARNED** *Practice the calculations of adjustments by completing Problem 5-9 at the end of the chapter.*

CHAPTER SUMMARY

The first five chapters of this book have outlined the procedures for the annual accounting procedures for a farm operation. These procedures culminate in accrual-adjusted financial statements that provide a report on the financial performance and financial position of the farm business. Market-based financial statements can be prepared to accompany the cost-based financial statements. The closing entries complete the journal entries for each year and prepare the accounts for the next year.

The following chapters provide details of various procedures needed to calculate the numbers shown on the financial statements. These chapters provide examples of disclosure notes and supplementary schedules to clarify the calculations. Various valuation methods for assets, liabilities, equity, revenues, and expenses are also presented.

PROBLEMS

5-1 ■ Using the answer from Problem 4-9 in Chapter 4, prepare an accrual-adjusted income statement.

5-2 ■ Use the unadjusted trial balance below and prepare a cash-basis income statement. Make a list of the differences between this income statement and the accrual-adjusted income statement in Problem 5-1.

UNADJUSTED TRIAL BALANCE		
Accounts	**Debits**	**Credits**
1000 Cash	$17,600.00	
1600 Machinery and Equipment	235,000.00	
1650 Office Equipment and Furniture	2,000.00	
1800 Land, Buildings and Improvements	650,000.00	
2310 Notes Payable Due within One Year		10,000.00
2400 Real Notes Payable—Non-Current		250,000.00
3100 Retained Capital		80,868.00
3130 Other Capital Contributions/Gifts/Inheritances		480,000.00
4000 Cash Crop Sales		120,000.00
6100 Wage Expense	12,000.00	
6110 Payroll Tax Expense	1,668.00	
6520 Herbicides, Pesticides	5,000.00	
6700 Insurance	5,200.00	
6710 Real Estate and Personal Property Taxes	7,200.00	
8200 Gains (Losses) on Sales of Farm Capital Assets	1,600.00	
9100 Income Tax Expense	3,600.00	
Totals	$940,868.00	$940,868.00

5-3 ■ Using the trial balance from Problem 4-9 in Chapter 4 and the answer from Problem 5-1 in this chapter, prepare an accrual-adjusted statement of owner equity. Then, using the unadjusted trial balance and the answer from Problem 5-2, prepare a cash-basis statement of owner equity. Make a list of the differences between the two statements.

5-4 ■ Using the answer from Problem 4-9 in Chapter 4 and the accrual-adjusted statement of owner equity from the answer in Problem 5-3 in this chapter, prepare an accrual-adjusted balance sheet. Using the unadjusted trial balance from Problem 5-2 and the cash-basis statement of owner equity from the answer in Problem 5-3, prepare a cash-basis balance sheet. Make a list of differences between the two balance sheets.

5-5 ▪ Using the cash account below, prepare a statement of cash flows. The payment on March 1 included $5,000 of interest paid.

1000 CASH					
Date	**Description**	**JE#**	**Debits**	**Credits**	**Balance**
Beginning Balance					$ 5,186
Jan. 30	Sale of stored grain		120,000		125,186
Feb. 1	Income tax payment			3,600	121,586
Mar. 1	To record interest and principal payment			50,000	71,586
Apr. 10	Sale of truck		5,000		76,586
Apr. 15	Purchase of new truck			32,000	44,586
June 1	Payment for herbicide			5,000	39,586
July 1	Payment of real estate and property taxes			7,200	32,386
Aug. 1	Payment of insurance premium for next 12 months			5,200	27,186
Oct. 1	Operating loan		10,000		37,186
Dec. 31	Employee paycheck			12,000	25,186
Apr. 10	FICA, FIT, FUTA, SUTA taxes paid			2,586	22,600

5-6 ▪ Using the information below, make adjustments to the cash-basis income statement in your answer to Problem 5-2, the cash-basis statement of owner equity in your answer to Problem 5-3, and the cash-basis balance sheet in your answer to Problem 5-4.

a. The value of the Farmers' raised crop (still in storage at the end of the year) was $105,500. They had no stored crop on hand at the beginning of the year.

b. The Farmers raised hay during the year and the value of the hay is $4,500 on hand on December 31. They had no raised feed on hand at the beginning of the year.

c. The Farmers purchased pelleted feed for $2,000 and salt and mineral blocks for $350 on December 15. All of it was still on hand at the end of the year. This purchase was not paid for by December 31. The Farmers had no purchased feed on hand at the beginning of the year.

d. The Farmers had no market livestock on hand at the beginning of the year but plan to purchase some feeder calves soon after the end of the year.

e. The Farmers purchased a one-year insurance policy on August 1 for $5,200. No prepaid expenses existed at the beginning of the year.

f. The Farmers determined that their unpaid bills at the end of the year are $2,350 for the feed that they purchased on December 15. They will not pay the bills before December 31. They had no bills at the beginning of the year.

g. The Farmers determined that the amount of current deferred taxes at the end of the year is $4,000. No current deferred taxes existed at the beginning of the year.

h. Depreciation expense amounted to $31,000.

5-7 ▪ Using the answers to Problem 5-6 and the market value information below, prepare market-based financial statements.

a. The Farmers purchased pelleted feed for $2,000 and salt and mineral blocks for $350 on December 15. All of it was still on hand at the end of the year. The market value of the pelleted feed was $2,500 at the end of the year. The market value of the salt and mineral blocks was similar to the purchase price of $350. The Farmers had no purchased feed on hand at the beginning of the year.

b. The market values of the Farmers' non-current assets were as follows at the end of the year:

	Cost	Market
Machinery and Equipment	$235,000.00	$240,000.00
Office Equipment and Furniture	2,000.00	2,000.00
Land, Buildings and Improvements	650,000.00	680,000.00

c. The Farmers determined that $5,290 is the amount of the second component of non-current deferred taxes at the end of the year. Non-current deferred taxes did not exist at the beginning of the year.

5-8 ▪ Prepare the closing entries from the trial balance in Problem 5-2.

5-9 ▪ Calculate the amount of the adjustments for each of the following situations.

a. The value of the Farmers' raised crop (still in storage at the end of the year) was $105,500. They had a stored crop on hand at the beginning of the year with a value of $108,000.

b. The Farmers raised hay during the year and the value of the hay is $4,500 on hand on December 31. The raised feed on hand at the beginning of the year had a value of $2,500.

c. The Farmers purchased pelleted feed for $2,000 and salt and mineral blocks for $350 on December 15. All of it was still on hand at the end of the year. This purchase was not paid for by December 31. The Farmers had purchased feed on hand at the beginning of the year with the following costs: pelleted feed, $1000; and grain, $500.

d. The Farmers purchased a one-year insurance policy on August 1 for $5,200. The amount of prepaid insurance at the beginning of the year was $1,800.

e. The Farmers determined that their unpaid bills at the end of the year are $2,350 for the feed that they purchased on December 15. They will not pay the bills before December 31. They had unpaid bills at the beginning of the year in the amount of $3,500.

f. The Farmers determined that the amount of current deferred taxes at the end of the year is $4,000. Current deferred taxes at the beginning of the year were $4,600.

CHAPTER 6

Revenue and Expense Measurements

I n Chapter 5, you learned the difference between an unadjusted trial balance and an adjusted trial balance; how to prepare financial statements; how to add adjustments to cash-basis financial statements to convert them into accrual-adjusted financial statements; how to prepare financial statements with market values; how to record closing journal entries and why it is necessary to record them; and how to calculate adjustments after the first year of operation.

You now have an understanding of accounting procedures. This is the first of four chapters that delve more deeply into the accounting principles and calculations involved in recording transactions and preparing financial statements. Performing calculations for financial statement items is often called **measurement**. You will learn about how to derive the numbers for transactions like those in Chapters 3, 4, and 5. Knowing how to perform these measurement procedures helps make financial statements understandable and reliable. The consistent use of these procedures allows for comparability from one year to the next or between farms.

This chapter introduces the rationale for the accounting procedures for revenues and expenses, including **revenue recognition**, and how it applies to the agricultural industry. You will learn how to account for government loans and how to calculate gains and losses, and will study the terms and methods related to depreciation. This chapter also teaches you how to calculate interest and Net Farm Income and, finally, how to define and recognize an extraordinary item.

Recall that you learned in Chapter 1 that one characteristic of useful financial statements is reliability for evaluating the financial position and financial performance of the farm operation. Reliable source documents can verify the information in the financial statements. The dollar amount of many revenue and expense items is easily determined from the source documents, but sometimes the amounts have to be calculated. In those cases, notes to the financial statements should verify the amount by disclosing information about the calculations. The disclosures are necessary because you can use more than one method of calculation for some items. Farm managers and accountants must be able to explain to outsiders the basis for the calculations for any specific operation. Without the knowledge of these methods, evaluations would be incomplete, and comparisons with previous years' financial statements or with other farm operations would most likely not be very useful.

At this point in your study of agricultural accounting, you should be familiar with the accounting procedures for such transactions as the selling of crops, market livestock and poultry, and livestock products (Chapter 3), and for changes in inventory and changes in accounts receivable in the accrual-adjusted system (Chapter 4). In this chapter's section on revenue recognition, you will learn about the rationale for reporting changes in inventories. Instead of just knowing *how* to report changes in inventory, you will know *why* to report these changes. Other topics in this chapter include reporting of revenue from government loans, and calculating gains and losses from selling and trading non-current assets.

You have also learned to record the cash expenses of a farm operation (Chapter 3). You are familiar with the accounting procedures for reporting year-end adjustments, such as changes in interest payable, changes in accounts payable, changes in prepaid insurance, changes in growing crops, and change in taxes payable in an accrual-adjusted system (Chapter 4). In this chapter, you will learn how to calculate depreciation and interest. You will also learn about the significance of the income statement item called **Net Farm Income**, used to evaluate the financial performance of the farm business. This chapter concludes with a discussion of a special topic called **extraordinary items**. Extraordinary items are unusual and infrequent circumstances that have a financial impact on the farm business.

REVENUE RECOGNITION

Learning Objective 1 ■ To identify the two main issues for accurate and complete reporting of revenue and the types of revenue that can occur.

In addition to verifying financial information from source documents, another feature of reliability in financial statements is accuracy and completeness (to the extent possible). Sometimes estimates have to be made. In those instances, accuracy may be compromised. Nevertheless, there should be a sound basis for the estimate. (We give an example of estimates in the topic of depreciation expense in this chapter.)

Two main issues concern the accurate and complete reporting of revenue. The first is the amount of revenue to record and the other is an issue called revenue recognition. In Chapter 1, you learned that revenue is defined as "money earned by the farm business from the production or sale of farm products." The concept of revenue recognition helps to explain how to determine when revenue has been earned.

Revenues for a farm operation can occur from several sources. Table 6-1 shows common sources of revenue for farm businesses. For some revenue transactions, the amount of revenue is the amount of the money from sales of farm products. In Chapter 3, you learned how to record cash sales. The check and other documents state the amount of each sale. Recall from Chapter 4 that in an accrual-adjusted system, gross revenue also includes "change" items that represent money earned from the production of farm products. Receipts or contracts can verify the amount of money owed to the farm business for calculating the change in accounts receivable.

TABLE 6-1 ■ *Sources of revenues for a farm operation.*

Sales

    Sales of crops
    Sales of feed
    Sales of market livestock and poultry
    Sales of breeding livestock (breeding farms)
    Sales of livestock products (for example, eggs, milk, wool)
    Changes in accounts receivable

Production

    Changes in crop inventories
    Changes in market livestock and poultry inventories
    Changes in breeding livestock inventory (breeding farms)
    Change in the value of raised breeding livestock due to changes in quantity
    Gains/losses due to changes in base values of raised breeding livestock

Other

    Payments from government programs
    Payments from crop insurance claims
    Gains/losses from the sale of culled breeding livestock
    Gains/losses from the sale of non-current farm assets
    Extraordinary gains/losses

Gross revenue also includes changes in inventories and changes in the values of raised breeding livestock. The value of crop and livestock inventory depends on the amount that was paid when purchased (verified by receipts) or the current market value (verified by quoting a reliable source of market prices and a reasonable estimate of weight or number of units). In Chapter 4, you learned how to report these changes. You will learn more about the accounting procedures for changes in raised breeding livestock in Chapter 8.

Other sources of revenue reported by the farm operation might include money received from government programs and crop insurance claims. Chapter 3 includes an example of the accounting for a payment from a crop insurance claim. In the next section, you will learn about reporting revenue from participating in a government loan program. Later in this chapter, you will read about gains and losses from the sales of culled breeding livestock or other farm assets, and from extraordinary occurrences.

To report revenue as accurately as possible, the farm accountant must know the amount of revenue earned and when it occurred. Revenue recognition concerns when to record and report revenue (see Table 6-2). For most retail and manufacturing businesses, the general rule is that revenue must be recognized (that is, recorded and reported) when the revenue has been earned and when the goods or services are exchanged for cash or accounts receivable. Those two conditions must occur for reporting revenue on the income statement.

Learning Objective 2 ■ To describe the general rule for revenue recognition and the exception to that rule for agricultural accounting.

- ■ In the first condition, the revenue has to be earned. That means that all of the activities creating the goods or services must be performed, that is, the product is ready for sale.
- ■ In the second condition, an exchange must take place between the seller and a buyer, that is, the goods or services are sold for cash or on account to a buyer.

TABLE 6-2 ■ *Revenue Recognition rules.*

General Rule: Two conditions must occur.

    1. Revenue must be earned (production is complete).

    2. Exchange takes place (sale has occurred).

Agriculture: First condition must occur.

    1. Revenue must be earned (production is complete).

        Revenue is reported if financial statements are prepared before the sale.

    2. Exchange takes place (sale has occurred).

        Revenue is reported when sale occurs also.

If a business produces a product, the sale price might not be determined until someone is willing to buy it. If a contract for producing a product exists between a seller and a buyer, but the seller has not produced the product yet, the activities have not yet been performed. In either case, revenue should not be recorded. The product has to be ready and it has to be sold. When these two conditions are met, a revenue transaction can be recorded and reported on the income statement, even if the cash has not yet been received.

Revenue recognition for farm products is somewhat different from the products and services of retail or manufacturing businesses (see Table 6-2). The first condition, that the revenue has to be earned, applies to farm products in the same way as the general rule described above. While raising or growing the grain, livestock, or livestock products, the activities to produce them are not complete. When the crops are harvested, the eggs are gathered, the wool is sheared, the milk is stored, the market livestock have reached an appropriate market weight, and so on, the farm products are ready to sell and the revenue has been earned. Until then, no revenue can be recorded. Only the expenses to produce the products are recorded as they occur.

The second condition in the general rule says that, in addition to the revenue being earned, a sale has to occur. However, revenue can be also recognized for farm products that are ready to sell if the financial statements need to be prepared before selling the product. This is an exception to the general rule above. As illustrated in Chapters 4 and 5, the income statement may report the revenue even though a sale has not yet taken place. Revenue is recognized from the production of farm products. The rationale is that for most farm products, there is little uncertainty about the market price and opportunity for the farm producer to sell the product, because active commodity markets exist.

In some cases, a sale is agreed to in advance of production. This type of situation occurs when a farm producer makes an agreement to deliver a crop after harvest, such as a **forward contract**, in which the producer is obligated to make delivery. In that case, the sale has already occurred but the product has not been produced, so no revenue is recorded until the product has been produced and is ready to be delivered. The first condition for revenue recognition still holds—that the revenue must be earned before any revenue can be recognized.

PRACTICE WHAT YOU HAVE LEARNED *Can you apply the concepts of revenue recognition so that you know when revenue needs to be recorded and reported? Work on Problem 6-1 at the end of the chapter to test your knowledge.*

GOVERNMENT LOANS

Government loan programs present a unique type of revenue situation. Under **government loan programs**, a farm producer is allowed to offer crops as collateral for a loan with the Commodity Credit Corporation (CCC) at a specified loan rate (principal) for up to nine months. When the cash is received, the farm accountant would record the transaction for the loan just as you learned in Chapter 3. Like other loans, the CCC loan should be shown as a current liability on the balance sheet (such as Notes Payable due within one year) until the loan is settled.

Learning Objective 3 ■ To record revenue from government loans.

If the financial statements need to be prepared before the loan is settled, the balance sheet reports the crop pledged as collateral as Crop Inventory. Table 6-3 outlines how you would value the inventory. The balance sheet reports the value of the inventory as either the market value of the crop (on the day that the balance sheet is prepared) or the loan rate, whichever is the higher number. If the market value is higher than the loan rate, the balance sheet reports the loan at the loan rate (as Notes Payable due within the next year) and the crop inventory at market value. According to the FFSC Guidelines, the balance sheet reports the difference between the market value and the loan rate as accrued interest (Interest Payable).

Unlike most other loans, the producer has the option of either repaying the loan or forfeiting the crop to the CCC. A producer who forfeits the crop keeps the cash proceeds from the loan and makes no payment to the CCC. No interest is paid. The cash proceeds are recorded as revenue when title to the crop passes to the CCC. If the producer forfeits the crop instead of paying back the loan with cash, the liability for the loan is removed from the balance sheet and the revenue from government programs is reported on the income statement. The journal entry to record the repayment and the revenue from government programs is as follows:

(Date)	2310 Notes Payable due within one year	XXX	
	4300 Proceeds from Government Programs		XXX

If, instead, the producer pays back the loan to the CCC, the payment is recorded in the same way as for any other loan repayment. The producer then sells the crop in the usual manner (or perhaps has already sold it) and the revenue from the sale is recorded as you learned in Chapter 3.

TABLE 6-3 ■ *Balance sheet reporting for government loans.*

Balance sheet prepared before loan is settled (loan rate > market value):

Inventory	Liabilities
Crop Inventory (at loan rate)	Notes Payable (at loan rate)

Balance sheet prepared before loan is settled (loan rate < market value):

Inventory	Liabilities
Crop Inventory (at market value)	Notes Payable (at loan rate)
	Interest Payable (market value minus loan rate)

> **Exercise 6-1** *Refer to the Farmers' financial statements in Appendix A. Their balance sheet lists Notes Payable within one year, Interest Payable, and Crop Inventory Raised for Sale. Can you tell from the balance sheet whether they participated in a government loan program?*
>
> *Answer: No. You would have to read the notes to the financial statements and look for an explanation about their liabilities to find out if any of their loans came from a government loan program. Another way to tell is to look at the statement of cash flows to read whether they received any loans from the government. You would look in the financing activities section for an item listed as "Proceeds received from government loan program." There is no such entry for the Farmers.*
>
> *Another question: If the Farmers paid back the government loan or forfeited the crop, how could you tell if the Farmers had participated in a government loan program? Answer: The cash proceeds would be listed as "Proceeds received from government loan program" on the income statement and in the operating activities section of the statement of cash flows.*

GAINS AND LOSSES

Learning Objective 4 ■ To calculate the gain or loss on a sale or trade-in of an asset and to recognize where these gains and losses are reported.

In addition to revenues and expenses, the income statement reports gains and losses. As you learned in Chapter 1, **gains** are financial benefits from various activities. **Losses** are the opposite of gains. Gains result in an increase in equity and losses result in decreases in equity. The gains and losses discussed in this section include gains and losses from the sales of culled breeding livestock and gains and losses from sales or trade-ins of other farm assets.

Sale of Farm Assets Other Than Inventory

Revenue is sometimes recorded (as a gain) when farm assets are sold. Sometimes these transactions could result in a loss instead of a gain. The amount of the gain or loss from these sales is the difference between the amount of cash received and the book value of the sold asset. **Book value** is the current market value of the asset or its original cost minus its accumulated depreciation. These amounts are shown on the last prepared balance sheet. The farm accountant performs two steps (outlined in Table 6-4) to determine the amount of the gain or loss to record after the sale of an asset.

TABLE 6-4 ■ *Calculating the Gain or Loss on a sale of farm assets.*

Step 1: Determine book value or market value of the asset from the last balance sheet that was prepared.

  Book value = original cost − accumulated depreciation

  Or

  Book value = current market value

Step 2: Compare book value with the amount of cash received. The difference is a gain or loss.

  Gain: When cash received is greater than book value.
  Loss: When cash received is less than book value.

In Chapter 3, the Farmers sold a bull for $900. They had paid $1,000 for the bull when they purchased it. When they sold it to the neighbor, the depreciation amounted to $300. In Step 1, the book value is the original cost minus the accumulated depreciation of the bull.

Step 1: Book value = $1,000 – 300
= $700

In Step 2, the $900 cash received is compared to the $700 book value, resulting in a $200 gain.

Step 2: Cash received – book value = $900 – 700
= $200 gain

When culled breeding livestock are sold, the gain or loss on the sale is included in the computation of Gross Revenue (see Chapter 1). If other non-current assets (land, machinery, or equipment, breeding livestock not sold as culled) are sold, the gain or loss is reported as Gains/Losses on the Sale of Farm Capital Assets and is added (or subtracted) from Net Farm Income from Operations.

Exercise 6-2 *Refer to the Farmers' financial statements in Appendix A. Did they report any gains or losses from the sale of farm assets? If so, what gains or losses did they report and how much were the amounts? Answer: Yes. The Farmers reported a loss of $250 on the sale of culled breeding livestock and a loss of $800 on the sale or trade-in of other farm assets.*

Trading in an Old Asset

Trade-ins are slightly more complicated. Gains or losses are determined by comparing the book value of the traded asset and the cash paid for the new asset with the value of the new asset purchased. Step 1 is the same as above. Table 6-5 outlines Step 2 for calculating the gain or loss on a trade-in.

TABLE 6-5 ▪ *Calculating the Gain or Loss on a trade-in of non-current farm assets.*

Step 1: Determine book value or market value of the asset from the last balance sheet that was prepared.

Book value = original cost – accumulated depreciation

Or

Book value = current market value

Step 2: Compare value of asset received (purchased) with the amount of cash paid and the book value of asset given up (traded in) in the transaction.

Gain: When value of asset received is greater than cash paid and asset given up.

Loss: When value of asset received is less than cash paid and asset given up.

In the Farmers' case of the trade-in of the old tractor for a new one (Chapter 3), the gain or loss is the difference between what they are giving in the deal (the book value of the old tractor and the cash they paid) and what they are receiving in return (a new tractor with a value of $50,000). They had purchased the old tractor for $30,000. It was fully depreciated at the time of the trade-in (accumulated depreciation = $25,000) and, therefore, had a book value of $5,000.

Step 1: Book value = $30,000 − 25,000
                         = $5,000

The dealer gave them a $4,000 trade-in allowance on the old tractor. The purchase price of the new tractor is $50,000. The $4,000 trade-in allowance means that the Farmers had to pay $46,000. Step 2 involves comparing what the Farmers received with what they gave in the deal. They received a $50,000 tractor and they gave a $5,000 tractor and $46,000 cash. The result is a $1,000 loss.

Step 2: New asset received − book value of asset given up − cash paid
                         = $50,000 − 5,000 − 46,000
                         = −$1,000 loss

When non-current assets are traded, the gain or loss is reported as Gains/Losses on the Sale of Farm Capital Assets and is added (or subtracted) from Net Farm Income from Operations.

> **PRACTICE WHAT YOU HAVE LEARNED**   *At this point, you should be able to calculate gains and losses on sales and trade-ins. Complete Problem 6-2 at the end of the chapter to test your knowledge.*

DEPRECIATION EXPENSE

Learning Objective 5 ■ To explain the meanings of the terms "depreciation expense," "accumulated depreciation," "book value," and "market value."

The recording and reporting of many operating expenses is straightforward because the receipt, bill, or canceled check verifies the amount. However, as you learned in Chapters 4 and 5, sometimes the farm accountant has to estimate the amount of an expense by comparing the unused amount at the end of the year with the amount that was unused at the beginning of the year. Estimates are also made for allocating the cost of non-current assets, such as breeding livestock, perennial crops, machinery and equipment, buildings and improvements, and leased assets. The cost is allocated because the asset is used for more than one period. Let us review some definitions and key concepts about cost allocation from Chapters 4 and 5.

- ■ **Depreciation** is the process of allocating the cost of non-current assets (except land) over the useful life of the assets (the years that the asset will be useful in the farm operation).
- ■ **Depreciation expense** is the amount of depreciation reported for a single year.

- **Accumulated depreciation** is a contra account to non-current assets that reports the total amount of the cost allocated since the asset was put into use.
- Land is not depreciated because land is not used up to the point where it cannot be used for some purpose. Land can last forever but other non-current assets last only for a limited period, so the cost of land does not have a time to use in the allocation. Unlike other assets, the farm operation can sell the land virtually in its original form at any time.
- **Salvage value** refers to the amount that a non-current asset can expected to be sold for at the end of its useful life.

Depreciation is not a method of determining the market value of non-current assets. **Market value** is the price that a buyer and a seller can agree upon in an exchange of an asset. Frequently, the market value and the book value are not the same for several reasons. The condition of the asset is one circumstance that affects its market value when it is appraised. Book value, on the other hand, is a computation that depends upon the original price of the asset and the method used to allocate its cost. Sometimes the allocation method tracks well with the deterioration of the asset, but not always.

Exercise 6-3 *Refer to the Farmers' financial statements in Appendix A. How much depreciation expense did the Farmers report and which financial statement reports it? How much accumulated depreciation did the Farmers report and which financial statement reports it? Answer: The income statement reports the depreciation expense and the balance sheet reports the accumulated depreciation. Both are reported for $22,150 in the Farmers' first year of operation.*

To determine the amount of depreciation to record in a given year, three items are required: the **original cost** of the asset (the purchase price), the **estimated useful life** (length of time in years that the asset will be used by the farm operation), and any salvage value. The farm accountant estimates the useful life of each asset based on experience in the use of farm assets. For example, a brood cow might have an estimated life of eight years. The farm manager might estimate that the cow could be sold for approximately $250 at the end of her useful life. The $250 represents the estimated salvage value. When the original cost is unknown, the balance sheet reports only the current market value with no depreciation.

The amount of depreciation expense depends on the method used to calculate depreciation. The simplest method of depreciation is the **straight-line method**, in which the amount of depreciation expense is the same amount each year. Depreciation is calculated in the following manner according to the straight-line method:

Depreciation expense = (Original cost – Salvage value) ÷ Estimated life

The original cost minus the salvage value represents the **base** or the amount of the cost allocated during the life of the asset. The base is divided by the estimated useful life to determine the amount of depreciation expense to record each year. Depreciation expense is recorded for the same amount every year that the asset is owned and used. Because of the ease of the calculation, many businesses, both farm and non-farm, use this method.

Learning Objective 6 ■ To calculate depreciation expense using the straight-line method and to describe how to report depreciation in the notes to the financial statements.

If Steve and Chris assume that the tractor that they purchased for $50,000 (journal entry (7) in Chapter 3) would last about 15 years and they decide to allocate the cost evenly over the estimated 15-year life of the tractor, the depreciation expense is $3,333 per year ($50,000 ÷ 15 years). They make similar calculations for each of their non-current assets. Steve and Chris estimate a 4-year life for the office furniture, a 5-year life for the truck, a 10-year life for the perennial crop, a 15-year life for the leased harvester, and a 25-year life for the farm buildings. The following schedule shows calculations for depreciation expense (using the straight-line method):

Asset	Cost	Life	Depreciation Expense per Year
Office Furniture	$ 1,000	4	$ 250
Buildings, Improvements	110,000	25	4,400
Tractor	50,000	15	3,333
Truck	15,000	5	3,000
Leased Harvester	100,000	15	6,667
Orchard	45,000	10	4,500
Total Annual Depreciation			$22,150

The income statement reports the total amount of depreciation expense for $22,150.

For the sake of simplicity, each of the assets in the example had no salvage value. However, most farm operators would agree that some of these assets would have salvage value. In the previous example, the culled breeding cow has salvage value because the cow can be sold for slaughter. If any of the assets listed above would have any salvage value, the salvage value should also be included in the table. The table should be included in the notes to financial statements to clarify for the readers how the amount of depreciation was calculated.

> **PRACTICE WHAT YOU HAVE LEARNED**   *To practice these calculations, complete Problem 6-3 at the end of the chapter.*

Learning Objective 7 ■ To explain the meaning of the terms "tax-based depreciation" and "accelerated methods of depreciation."

The straight-line method is simple to use, but there are other depreciation methods. The amount of depreciation expense that is reported could vary substantially from one farm to another due to different calculations for depreciation. When an outside party compares one income statement with another that uses a different depreciation method, some knowledge of the calculations allows the outside party to analyze this difference accordingly. Although straight-line depreciation is permitted for tax purposes, many businesses are advised to use **accelerated methods of depreciation**. Accelerated methods calculate a greater amount of depreciation during the early years of the asset's life and less in the later years than the straight-line method would. The amount of depreciation declines each year, so the depreciation expense must be calculated each year. For that reason, accelerated methods create more complexity in record keeping.

Some farm businesses might use **tax-based depreciation**, which is calculated according to tax rules. The calculations are not presented in this book primarily because tax accountants use this method. Tax-based depreciation is an accelerated method, and more complicated than straight-line depreciation. Even so, many farm accountants might use it to make depreciation expense on the farm financial statements consistent with the amount shown on the tax return. The advantage is that farm accountants do not need to calculate depreciation expense; they simply ask the tax accountant to perform the calculations as if for the tax return, and then they record the amount of depreciation calculated by the tax accountant. However, the financial statements can be prepared using the straight-line (or some other) method while the tax return can be prepared using a tax-based method. The notes to the financial statements should disclose the method of depreciation method used, along with schedules (tables) showing the calculations.

In the straight-line method, depreciation expense is an allocation of the asset's cost evenly throughout the asset's expected useful life. The amount of depreciation expense is the same every year that the asset is used, unless the asset's useful life or salvage value is re-estimated for some reason. Learning Objective 6 in this chapter discusses these changes.

Learning Objective 8 ■ To apply various depreciation methods to calculate depreciation expense.

The straight-line method is easy to apply because it only has to be calculated once, when the asset is purchased. The farm accountant records the same amount of depreciation expense each year. Because of its simplicity, this method is the most popular depreciation method for financial reporting. The method is quite useful for buildings and improvements, perennial crops, breeding livestock, and some machinery and equipment because of regular use during its life.

Some machinery and equipment might be used more in some years than others. In that case, an alternative method, called the **activity method** of depreciation, is appropriate for calculating depreciation expense. To use this method, the farm accountant has to decide on a measure of activity. Because many assets do not have such a measure, this method can be useful only for assets such as vehicles and tractors. Vehicles have an odometer that measures miles and tractors have a similar device that measures hours of use.

■ To calculate depreciation expense under the activity method, the farm accountant has to estimate the total amount of miles or hours of operation for the vehicle or equipment during its entire useful life.

■ Divide the base by the total estimated hours or miles and then multiply by the amount of hours or miles operated in the current year.

Depreciation expense = (Base ÷ total estimated hours or miles) × actual hours or miles this year.

Activity depreciation expense has to be recalculated each year because the number of actual hours or miles will vary from year to year. However, the activity method is more accurate than the straight-line method for some assets because of the varying amounts of use. It is useful for those assets for which the amount of activity can be measured.

The Farmers estimated that their pickup truck has a useful life of 100,000 miles before they will trade it for a new truck. If they purchased it for $15,000 and they estimate that the trade-in (market) value will be approximately $3,000, then the base is $12,000. If they drove the truck 22,000 miles in the year 20X1, under the activity method they would divide the $12,000 base by the total estimated miles of 100,000 and then multiply that by the 22,000 miles that they drove to determine depreciation expense:

Depreciation expense = ($12,000 ÷ 100,000) × 22,000 = $2,640

If the Farmers chose to use the straight-line method and they estimated the useful life of the truck to be five years, then they would calculate depreciation expense as follows:

Depreciation expense = $12,000 ÷ 5 = $2,400

Another group of depreciation methods are known as **accelerated depreciation methods**. In these methods, depreciation expense is quite high the first year and then declines each year. The reason for using one of the accelerated methods is that for some assets, the repair expenses are low when the asset is new and they increase as the asset gets older. To offset the increasing repairs, lower amounts of depreciation expense are recorded as the asset gets older. When the asset is new and repairs are low, the depreciation expense is higher. As a result, net income is more "level" or "smoothed out." These methods are appropriate for assets that require high maintenance costs as the asset gets older, such as machinery or land improvements. The farm accountant has to decide whether these methods would be worth the extra computations required to apply them.

One of the accelerated methods is called the **sum-of-years-digits method**.

- In this method, the farm accountant multiplies the base by a fraction each year to compute depreciation expense. The fraction is smaller and smaller with each consecutive year, so depreciation expense declines each year. The fraction is based on a mathematical formula that makes this possible.

  Depreciation expense = Base × Fraction

- The denominator is the sum of the digits of the estimated years in the life of the asset. A shortcut way to calculate the denominator is the formula $[n \times (n + 1)] \div 2$, where "n" is the number of years in the asset's life.
- The numerator in the fraction is the number of years remaining in the asset's life. This number declines every year, so the fraction declines every year.

The Farmers' truck with a 5-year life would have the denominator calculated in this way:

1 + 2 + 3 + 4 + 5 = 15

The formula yields the same result as calculated above:

[5 × (5 + 1)] ÷ 2 =
[5 × 6] ÷ 2 =
30 ÷ 2 = 15

The numerator in the first year that the Farmers owned the truck is 5. The second year, the numerator is 4, and so on. A depreciation schedule showing how the Farmers calculated depreciation appears in the disclosure notes as follows:

Depreciation expense = Basis × Fraction
Year 1: $12,000 × (5/15) = $4,000
Year 2: 12,000 × (4/15) =   3,200
Year 3: 12,000 × (3/15) =   2,400
Year 4: 12,000 × (2/15) =   1,600
Year 5: 12,000 × (1/15) =   800
Total $12,000

At the end of Year 5, the truck will be completely depreciated and they will record no more depreciation expense for the truck even if they keep using it.

Another type of accelerated method is called the **declining-balance method**.

- In this method, the farm accountant multiplies a multiple of the straight-line rate by the book value of the asset (not the base) to calculate depreciation expense. The straight-line rate is 1 ÷ the number of years in the asset's life (n).

    Straight-line rate = 1 ÷ n

    Depreciation expense = Book value × Multiple × Straight-line rate

- In the first year, the book value is the original cost.
- The second year (and each subsequent year), the book value is the original cost minus all of the depreciation recorded in previous years (accumulated depreciation).
- A multiple of the straight-line rate is some number multiplied times 1 ÷ n. If the multiple were 2, then the straight-line rate of 1 ÷ n would be multiplied times 2. Any multiple can be used, but 1.5 and 2 are common.

    Multiple × Straight-line rate = 2 × (1 ÷ n)

- When the book value is close to the salvage value, the amount of depreciation is only the amount that will bring the book value to the amount of the salvage value.

The Farmers' truck has a 5-year life and an estimated salvage value of $3,000. The straight-line rate for the Farmers' truck is 1 ÷ 5. If the Farmers decide to use a multiple of 2, then 2 × (1 ÷ 5) is multiplied by the book value of the truck. In the first year, the book value is the original cost of $15,000. The depreciation expense recorded the first year is:

    Depreciation expense = $15,000 × 2 × (1 ÷ 5) = $6,000

The next year the book value is:

    Cost − Accumulated depreciation
    $15,000 − 6,000 = $9,000

The depreciation expense recorded the second year is:

    Depreciation expense = $9,000 × 2 × (1 ÷ 5) = $3,600

The book value for the third year is:

$15,000 − 6,000 − 3,600 = $5,400

The depreciation expense for the third year is:

Depreciation expense = $5,400 × 2 × (1 ÷ 5) = $2,160

The book value for the fourth year is:

$15,000 − 6,000 − 3,600 − 2,160 = $3,240

The book value is only $240 above the salvage value, so depreciation expense for the fourth year is:

Depreciation expense = $3,240 − 3,000 = $240

The book value is:

$15,000 − 6,000 − 3,600 − 2,160 − 240 = $3,000

You can see the complexity of these methods in these examples. Because of this complexity, the accelerated methods are less commonly used than the straight-line method.

Table 6-6 compares the depreciation expense under the straight-line, sum-of-years-digits, and declining balance methods for the Farmers' truck. The Farmers would have to decide which of the methods discussed would be the most suitable for each of their non-current assets, and record depreciation expense accordingly. They also have to decide if it would be more suitable, instead, for them to use the depreciation amounts calculated by the tax accountant.

Table 6-6 illustrates the nature of the accelerated methods. Depreciation expense is less each year under the accelerated methods. In Years 1 and 2, the depreciation expense is greater under the accelerated methods than it is under the straight-line method. This situation reverses in later years when the straight-line depreciation expense is greater than accelerated depreciation expense.

PRACTICE WHAT YOU HAVE LEARNED *Practice the depreciation methods by completing Problem 6-4 at the end of the chapter.*

TABLE 6-6 ■ *Comparison of different depreciation methods.*

	Straight-Line	Sum-of-Years-Digits	Declining Balance
Year 1:	$2,400	$4,000	$6,000
Year 2:	2,400	3,200	3,600
Year 3:	2,400	2,400	2,160
Year 4:	2,400	1,600	240
Year 5:	2,400	800	0
	$12,000	$12,000	$12,000

Partial Periods

A partial period concerns the depreciation expense for part of the year. The farm accountant generally calculates depreciation expense for an entire year (12 months). If an asset was purchased at the beginning of the year, then the 12-month calculation is appropriate because the farm operation owns and uses the asset for approximately 12 months in the first year. However, most assets are not purchased right at the beginning of the year, so they are owned for only part of the first year. The farm accountant can choose to disregard the partial year, or may record only part of the 12-month depreciation in the year that the asset was purchased (see Figure 6-1).

Learning Objective 9 ■ To apply depreciation methods to partial periods and revised estimates.

- ■ To calculate depreciation expense for a partial year, the farm accountant multiplies the amount of depreciation expense for 12 months by the fraction of the year that the asset was owned.

For example, if the Farmers purchased the truck on October 1, 20X1, they owned it for 3 months or 3/12ths of the year 20X1. Therefore, they record only 3/12ths of the depreciation in the year 20X1. Under the straight-line method, this approach is quite easy to apply. The 12-month amount of depreciation is multiplied by 3/12ths.

Depreciation expense (for 20X1) = ($12,000 ÷ 5) × 3/12 = $600

For each of the next four years, the amount of recorded depreciation is $2,400, the full 12-month amount. In the last year of ownership, the amount of depreciation is the amount left to depreciate. At the end of the fifth year, the accumulated depreciation is $10,200.

20X1: Depreciation expense =	($12,000 ÷ 5) × 3/12 = $	600
20X2: Depreciation expense =	$12,000 ÷ 5 =	2,400
20X3: Depreciation expense =	$12,000 ÷ 5 =	2,400
20X4: Depreciation expense =	$12,000 ÷ 5 =	2,400
20X5: Depreciation expense =	$12,000 ÷ 5 =	2,400
Accumulated depreciation		$10,200

The amount left to depreciate is $12,000 – 10,200 = $1,800. Therefore, the amount of depreciation expense for the year 20X6 is $1,800.

FIGURE 6-1 ■ *Partial year depreciation.*

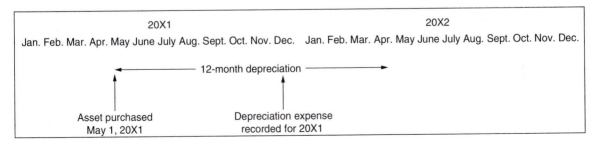

20X1	20X2
Jan. Feb. Mar. Apr. May June July Aug. Sept. Oct. Nov. Dec.	Jan. Feb. Mar. Apr. May June July Aug. Sept. Oct. Nov. Dec.

Asset purchased May 1, 20X1 — 12-month depreciation — Depreciation expense recorded for 20X1

This allocation for partial periods can also be performed for the sum-of-years-digits and declining-balance methods. Appendix J contains an example of calculating partial year's depreciation expense using accelerated methods. The activity method of depreciation does not calculate partial year depreciation expense because that method is not based on time, but rather on use. If an asset is used for only part of a year, depreciation expense is based on only the amount of miles or hours measured for that partial year.

> **PRACTICE WHAT YOU HAVE LEARNED** *Calculate partial year depreciation expense in Problem 6-5 at the end of the chapter.*

Revised Estimates

Revised estimates concern re-estimating the useful life of an asset or its salvage value. The farm accountant estimates the length of the useful life and the salvage value when the asset is purchased. Sometimes, as the years go by, these estimates are no longer valid. For example, if a machine is damaged and cannot be used as long as originally thought, the depreciation expense should be recalculated to reflect the remaining number of years in the asset's useful life. If the price of culled breeding livestock changes substantially, the farm accountant should recalculate depreciation expense to reflect the change in salvage value. Only the book value of the asset should be used to recalculate depreciation expense, not the base or original cost.

To recalculate depreciation expense when the useful life or salvage value is re-estimated, the farm accountant should apply the following procedures.

- First, determine the book value of the asset. The amount for accumulated depreciation depends on the depreciation method used and how much time has passed since the asset was purchased.

  Book value = Cost − Accumulated depreciation

- Next, subtract the estimated salvage value from the book value to calculate a new base. If you have re-estimated the salvage value, then subtract the new, re-estimated salvage value, not the old salvage value. If the salvage value is the same as before, then the old salvage value is subtracted.

  New base = Book value − Estimated salvage value

- Depreciation expense is calculated using the new base and the remaining years that the asset is expected to be useful (if using the straight-line method).

  Depreciation expense = New base ÷ Estimated remaining years of useful life

Suppose that the Farmers' truck wears out faster than they expected. In 20X4, they decided that they would not wait until 20X6 to trade it in and purchase a new truck. Rather, they will purchase a new truck in 20X5. Depreciation expense should be recalculated in 20X4 for the years 20X4 and 20X5.

At the end of 20X3, the book value for the truck is determined by subtracting the depreciation expense from 20X1, 20X2, and 20X3 (the accumulated depreciation) from the original cost. The Farmers used the straight-line method for depreciation. Just to show you the difference between the straight-line and the accelerated methods, the book values for the Farmers' truck for each of these methods are calculated below (using partial year depreciation in the first year):

	Straight-Line	Sum-of-Years-Digits	Declining Balance
Original cost:	$15,000	$15,000	$15,000
Less accumulated depreciation:			
20X1:	600	1,000	1,500
20X2:	2,400	3,800	5,400
20X3:	2,400	3,000*	3,240**
Book value:	$ 9,600	$ 7,200	$ 4,860

* $3,000 = (266.67 \times 9) + [(2400 \div 12) \times 3]$ (See Appendix J).
** $3,240 = (300 \times 9) + [(2400 \div 12) \times 3]$ (See Appendix J).

The Farmers are assuming that the salvage value will be the same as they originally thought. They subtract the old salvage value of $3,000 from the book value to calculate the new base. For each of the methods, the new base is calculated as follows:

	Straight-Line	Sum-of-Years-Digits	Declining Balance
Book value:	$ 9,600	$ 7,200	$ 4,860
Old salvage value:	− 3,000	− 3,000	− 3,000
New basis:	$ 6,600	$ 4,200	$ 1,860

Under the straight-line method, they calculate the revised depreciation expense by dividing the new base by the two remaining years (20X4 and 20X5):

Depreciation expense (for 20X4 and 20X5) = $6,600 ÷ 2 = $3,300

Under the sum-of-years-digits method, a new fraction is calculated by adding 1 + 2 for the denominator, and using 2 as the numerator for the year 20X4 and using 1 as the numerator for the year 20X5:

Depreciation expense (for 20X4) = $4,200 × 2/3 = $2,800
Depreciation expense (for 20X5) = $4,200 × 1/3 = $1,400

Under the declining balance method, the straight-line rate is 1 ÷ 2, so twice the straight-line rate is 2 × (1 ÷ 2) or 1. The entire new base is depreciated in 20X4.

Depreciation expense (for 20X4) = $4,860 − 3,000 = $1,860

The truck is fully depreciated down to the salvage value under each of these methods by the end of the year 20X5.

	Straight-Line	Sum-of-Years-Digits	Declining Balance
Book value at Jan. 1, 20X4:	$ 9,600	$ 7,200	$4,860
Depreciation:			
20X4:	3,300	2,800	1,860
20X5:	3,300	1,400	0
Book value at Dec. 31, 20X5:	$ 3,000	$ 3,000	$3,000

Answers: A. Not extraordinary. This type of expense is an ordinary occurrence for a farm operation even though it might not happen very often. B. Depends on the location. Some areas of the country do not experience tornadoes. If the tornado occurred in Kansas, however, this would probably not be considered extraordinary. C. Not extraordinary. The loss from the hailstorm is an unusual occurrence. However, a farmer who insured the crop must have anticipated that hailstorms and crop damage are frequent enough that they are not considered extraordinary. D. This payment is a gain. Whether or not it is considered extraordinary depends on whether or not it will reoccur and how unusual it is for the area in which the farm is located. It also depends on how much money was received in the payment. A small amount would not be extraordinary.

CHAPTER SUMMARY

Topics in this chapter include concepts and measurement issues regarding farm revenues and expenses. Revenue recognition concerns when revenue should be recognized and recorded. Accounting for the special issue of revenue from participation in government loan programs is also discussed. The calculation of various gains and losses are based on book values and the amount received on a sale or trade-in. Measurement issues also include the calculation of depreciation, presented in this chapter with the widely used method called straight-line depreciation. Accelerated depreciation methods, partial periods, and revised estimates are also discussed. The measurement of interest expense involves the factor of time and interest rates. Other measurement issues covered in this chapter include Net Farm Income and extraordinary items. Although other issues involving revenues and expenses may arise in a farm business, these topics are common for many farm operations.

PROBLEMS

6-1 ■ For each of the cases below, indicate whether or not the conditions for revenue recognition have been met.

a. A wheat crop has been planted.

b. Calves are weaned.

c. Apple crop has been harvested.

d. Corn crop has been sprayed.

e. All feeder pigs have reached market weight.

f. Dairy cows have been milked.

g. Sheep have been shorn.

h. Hay crop has been sold.

i. A contract to deliver soybean crop has been signed.

6-2 ■ For each of the cases below, calculate the amount of the gain or loss.

a. A truck with a cost of $32,000 and accumulated depreciation of $15,000 is sold for $15,000.

b. A truck with a cost of $32,000 and accumulated depreciation of $15,000 is sold for $18,000.

c. A truck with a cost of $32,000 and accumulated depreciation of $15,000 is traded for a new truck. The dealer is giving a trade-in allowance of $15,000 on the old truck and the farmer has to pay $18,000 to buy the new truck.

d. A truck with a cost of $32,000 and accumulated depreciation of $15,000 is traded for a new truck. The dealer is giving a trade-in allowance of $18,000 on the old truck and the farmer has to pay $15,000 to buy the new truck.

e. Cows with base values of $3,000 were sold for $3,500.

f. Cows with base values of $3,000 were sold for $2,700.

g. The last time land was appraised, a parcel of land was reported with a market value of $60,000. The land was sold for $56,000.

h. The last time land was appraised, a parcel of land was reported with a market value of $60,000. The land was sold for $65,000.

6-3 ▪ Using the straight-line method, calculate the amount of depreciation expense for each asset and prepare a schedule that would be included in the notes to the financial statements.

Asset	Cost	Salvage Value	Estimated life
Truck	$ 32,000	$5,000	6 years
Tractor	45,000	3,000	14 years
Corn planter	35,000	0	15 years
Barn	100,000	6,000	30 years
Fences	25,000	0	20 years
Cattle squeeze chute	5,000	500	15 years
Computer	3,000	300	5 years

6-4 ▪ a. Calculate the amount of depreciation expense for one year for the buildings and improvements listed below if the buildings have an average estimated life of 20 years and the improvements have an average estimated life of 10 years with no salvage value. Use the straight-line method.

Buildings: Cost = $125,000

Improvements: Cost = $62,500

b. Use the activity method and calculate depreciation expense on the Farmers' truck for 20X2 and 20X3. They drove the truck 24,500 miles in 20X2 and 22,600 miles in 20X3.

6-5 ▪ Suppose that the Farmers purchased the buildings and improvements in Problem 6-4 on April 1. Calculate the amount of partial year depreciation expense for the buildings and improvements for the first year.

6-6 ▪ Suppose that the Farmers decided in 20X4 that they would keep the truck a year longer than they originally thought and will trade it and buy a new truck in 20X7. The salvage value will be $2,000 in 20X7. Calculate the revised amount of depreciation expense that they should record in 20X4, 20X5, and 20X6.

6-7 ▪ Prepare an amortization schedule using the following information: $75,000 principal, 10 percent annual interest rate, five annual payments of $19,784.65 each.

6-8 ■ Calculate Accrual Adjusted Net Farm Income using the appropriate data below.

Gain Due to Changes in General Base Values of Breeding Livestock: $5,000
Income Tax Expense: $5,400
Loss on the Sale of Farm Capital Assets: $1,600
Net Farm Income from Operations: $175,643
Other Revenue: $300
Other Expenses: $160

CHAPTER 7

Asset Valuation, Part 1—Current Assets

<table>
<tr><td colspan="4">Key Terms</td></tr>
<tr>
<td>Allowance method</td>
<td>Current assets</td>
<td>Errors</td>
<td>Net realizable value method</td>
</tr>
<tr>
<td>Bad debts or uncollectible
accounts</td>
<td>Current market value</td>
<td>First-In-First-Out (FIFO)</td>
<td>Orderly liquidation</td>
</tr>
<tr>
<td></td>
<td>Current market value method</td>
<td>Historical cost</td>
<td>Outstanding checks</td>
</tr>
<tr>
<td>Bank charges</td>
<td>Deposits in transit</td>
<td>Historical cost method</td>
<td>Relevance</td>
</tr>
<tr>
<td>Bank credits</td>
<td>Direct write-off method</td>
<td>Last-In-First-Out (LIFO)</td>
<td>Reliability</td>
</tr>
<tr>
<td>Bank reconciliation schedule</td>
<td>Discounted cash flow method</td>
<td>Net realizable value (NRV)</td>
<td>Weighted average method</td>
</tr>
<tr>
<td>Bank statement</td>
<td></td>
<td></td>
<td></td>
</tr>
</table>

In Chapter 6 you learned about some of the major topics concerning the income statement, including revenue recognition, accounting for government loans, how to calculate and report various gains and losses, how to calculate depreciation expense using the straight-line method, how to calculate interest expense, how to determine Accrual Adjusted Net Farm Income, and how to report extraordinary items.

Now our attention turns to the balance sheet. This chapter focuses on current assets. In Chapter 1, you learned that **current assets** are those assets expected to earn money or produce other benefits within one year. Examples of current assets are cash, accounts receivable, inventory, and prepaid expenses.

In Chapter 6, you learned that sometimes farm accountants obtain the amount to record for the revenue and expense transactions from the source documents, but sometimes they have to calculate or estimate the amount. In Chapters 3, 4, and 5, you learned how to record various transactions that involve current assets. In those examples, the amount to record was "given." However, like revenues and expenses, sometimes the value of current assets must be calculated or estimated. GAAP provides the rules for these calculations. The FFSC acknowledges several methods for calculating or estimating the value of assets. The recommendations for the value of assets in the FFSC Guidelines correspond to GAAP in some ways and depart in other ways. The suitability of the method depends on the nature of the asset. This chapter discusses these recommendations for current assets of farm businesses.

The topics in this chapter include valuation methods, bank reconciliations, accounts receivable and bad debt expense, inventory valuation, and information disclosed in the notes and schedules that accompany the financial statements.

VALUATION METHODS

According to GAAP, assets are generally valued at historical cost. **Historical cost** refers to the original cost of the asset. Historical cost is a reliable basis for valuation because it exists as a historical event, verified if purchase records exist. An alternative value is the **current market value**, usually defined as the price obtained in the marketplace between a willing buyer and a willing seller. GAAP discourages the use of current market values for most assets because market values are sometimes difficult to obtain or verify. If a readily accessible market does not exist for some assets, the validity of the market values might be questioned.

GAAP permits an exception to the market value rule for marketable securities because a readily accessible market value exists in the marketplace. Adjustments for marketable securities are performed at the end of the year, similar to the market value adjustments that you learned about in Chapter 5. Another notable exception is the value of inventory. GAAP requires that inventory be valued at the end of the year at the lower of the original cost or the current market value, but also allows primary market values for some agricultural inventories.

The FFSC Guidelines allow market values as an acceptable alternative to historical cost for all assets. Agricultural lenders prefer market values because market values serve as a better basis for making loans than historical costs. In addition, historical costs might not be available for some assets due to the lack of records. Furthermore, the value of farm assets appreciated greatly during the 1980s, and historical costs for those assets are not comparable to their current market values.

Although several valuation methods exist, the FFSC recognizes the following four commonly used methods. When reporting market values on the balance sheet, the notes to the financial statements should disclose the source of the market value estimates.

- The **historical cost method** applies the amount paid for the asset. The value shown on the balance sheet is the book value, which is the original cost adjusted for depreciation.
- The **current market value method** applies an estimate of the amount of cash that could be obtained from selling the asset in an orderly liquidation. An **orderly liquidation** is an exchange in which neither the buyer nor seller are forced to engage in the transaction. In a forced liquidation (such as bankruptcy), the seller often must sell an item in a relatively short period, and might not have time to acquire the best price. For some farm assets, the farm manager might have to hire a qualified appraiser to establish a current market value because a readily accessible market value may not exist.
- The **net realizable value method** applies the selling price of the asset, minus any direct selling costs. This method is particularly relevant for raised crops and livestock.
- The **discounted cash flow method** applies an estimate of cash inflows and outflows from using the asset over the estimated life of the asset. This method works for certain types of lease arrangements. (Chapter 9 shows you how to account for leases.)

When valuing assets, the valuation method must meet tests of relevance and reliability. **Relevance** is a GAAP term that refers to the capacity of information to make a difference to parties who make decisions based on the information. If the valuation method is relevant, it is useful. **Reliability** means that the information truly represents what it says it represents, that it is verifiable by another party, and that it is unbiased and free of errors. If the valuation method is reliable, it can represent the value of the asset fairly and the value is verifiable. The valuation method to use will depend on the nature of the asset being valued, which you will learn about in this chapter.

> **PRACTICE WHAT YOU HAVE LEARNED** | *To become familiar with these terms, complete the matching exercise in Problem 7-1 at the end of the chapter.*

CASH

Cash is an asset that does not require "valuation" per se, because its value is quite apparent. However, the cash account of a business requires continuous management due to the large number of cash transactions and the importance of maintaining the account for tax return preparation. The farm operation manages its cash with checking accounts, savings accounts, and perhaps also with money market funds, certificates of deposit, or other investments. Just as you keep track of deposits to your checking account and the checks you write, a farm business must also do so.

Learning Objective 2 ■ To reconcile the bank statement with the cash account.

One of the primary cash-management tasks is reconciling the bank account with the ledger account. The farm business receives a **bank statement**, which reports all of the cash transactions that have passed through the bank. The farm accountant must make the cash account in the ledger agree with the bank statement. This process is necessary because the cash balance on the bank statement is seldom the same amount as the cash balance in the ledger account. Differences between the ledger balance and the bank balance might occur for several reasons. Typical differences include deposits in transit, outstanding checks, bank charges, bank credits, and errors.

- **Deposits in transit** are cash amounts deposited but missing from the bank statement because they have not yet cleared the bank.
- **Outstanding checks** are checks written but not cleared by the bank because the recipient has not yet deposited the check or because the bank did not process the check before it issued the bank statement.
- **Bank charges** are charges by the bank against the farm bank account for service charges, printing checks, nonsufficient funds, electronic fund transfers, or other services performed by the bank for the farm business bank accounts.
- **Bank credits** are collections or deposits made by the bank on behalf of the farm account, such as interest earned on interest-bearing accounts, and money collected and deposited into the account directly at the bank.
- **Errors** can occur for several reasons, by either the bank or the farm accountant. To check for errors, the farm accountant verifies the amounts of deposits and checks by comparing each to the amounts recorded in the ledger account.

Reconciliation can detect errors in either the bank balance or the ledger balance and can alert the farm accountant to cash transactions that occurred at the bank without the accountant's knowledge. A **bank reconciliation schedule** is a table that explains the differences between the bank statement and the ledger account. The schedule consists of two parts: the reconciliation from the bank statement to the ledger account and the reconciliation from the ledger account to the bank statement.

The first step in reconciliation is to compare the items listed in the bank statement with those in the ledger account and check off each item listed in both. The process begins by comparing debits in the ledger with deposits and credits on the bank statement, followed by comparing credits in the ledger with checks and debits on the bank statement. The dates, the dollar amounts, and the check numbers are examined for accuracy (see Figure 7-1).

The next step is to classify items not checked off on either the bank statement or the ledger account as deposits in transit, outstanding checks, bank charges, bank credits, and errors (see Figure 7-2).

FIGURE 7-1 ■ *Checking Off Items on Bank Statement and Cash Account.*

Ledger:

Cash

Date	Debits	Credits	Balance
	√8,000		8,000
		√680	7,320
		1,000	6,320

Bank statement:
Deposits	√$8,000
Checks	√680
Bank fee	15
Balance	$7,305

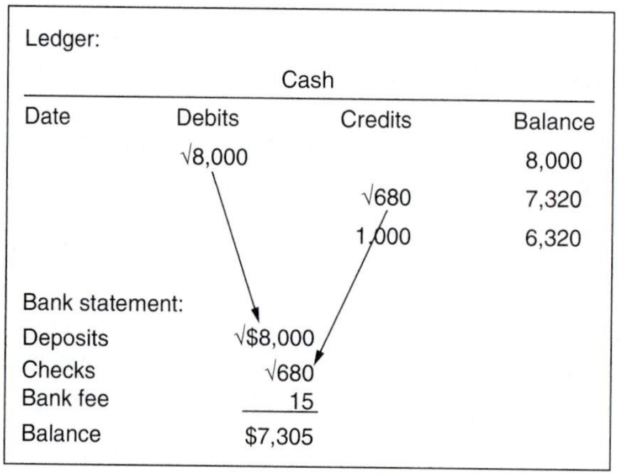

FIGURE 7-2 ■ *Classifying Unchecked Items on Bank Statement and Cash Account.*

Ledger:

Cash

Date	Debits	Credits	
	√8,000		
		√680	
		1,000	→ Outstanding check

Bank statement:
Deposits	√$8,000	
Checks	√680	
Bank fee	15	→ Bank charges

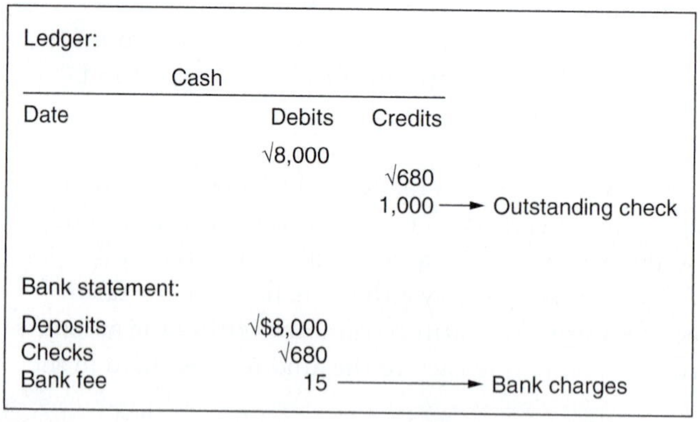

FIGURE 7-3 ■ *Format for Bank Reconciliation Schedule.*

```
                    Bank Reconciliation
          For the period ending (month, day, year)

Balance per bank statement                           $xxxxxx
     Deposits in transit:
              (date)                  $xxx               xxx
     Outstanding checks:
              (check number)          $xxx
              (check number)           xxx
              Total                                     xxxx
     Bank errors:                                        xx
Corrected Cash Balance                               $xxxxxx

Balance per books                                    $xxxxxx
     Bank charges:
              (date)                  $  xx
              Total                                     (xx)
     Bank credits:                                       xx
     Book errors:                                        xx
Corrected Cash Balance                               $xxxxxx
```

Next, the farm accountant prepares the bank reconciliation schedule. Figure 7-3 shows the format for the bank reconciliation schedule.

- The first section, beginning with balance per bank statement, reconciles the bank balance to the ledger account. The "balance per bank statement" is the cash balance at the bottom of the bank statement.
- The items listed below the balance per bank statement consist of the unchecked items in the ledger account. These items appear in the ledger account, but do not appear on the bank statement.
- The corrected cash balance is the amount that would appear on the bank statement and in the ledger account if all of the items appeared in both places.
- The second section, beginning with balance per books, reconciles the ledger account balance with the balance on the bank statement. The "balance per books" is the balance of the cash account in the ledger.
- The items listed below the balance per books consist of the unchecked items on the bank statement. These items appear on the bank statement, but not in the ledger account. The farm cash account must have a record of these items to show the correct cash balance in the ledger.
- The bank reconciliation is complete when the correct cash balance is the same in both sections of the schedule. If these two numbers are not equal, something has been overlooked and the items on the bank statement and in the ledger account must be rechecked. The process continues until the correct cash balances match.
- The final step is to record the items in the second section in journal entries. Items listed in the first section of the schedule (unchecked items in the ledger account) are not recorded because they already exist in the cash account. After posting all journal entries, the cash account balance should be equal to the corrected cash balance on the bank reconciliation.

```
(Date)          Bank Charges              15
                     Cash                            15
```

The Farmers completed bank reconciliations at the end of January and at the end of February 20X1. Now at the end of March, they will go through the procedures for the March bank statement. At the end of March, the Farmers' cash account in the ledger has the following entries. The items that cleared the bank in January and February have check marks placed beside them.

1000 Cash

Date	Description	JE#	Debits	Credits	Balance
Beginning Balance					0
Jan. 2	Set up farm bank account	(1)	√20,000		20,000
Feb. 1	Income tax payment	(31)		2,160	17,840
Feb. 12	Transfer from personal checking account	(3)	√100		17,940
Mar. 1	Money borrowed to purchase land	(6)	120,000		137,940
Mar. 1	Money borrowed to purchase tractor	(7)	50,000		187,940
Mar. 1	Purchase of land	(9)		120,000	67,940
Mar. 1	Purchase of tractor	(11)		46,000	21,940
Mar. 10	Sale of bull to neighbor	(10)	900		22,840

As you can see, the payment for income taxes had not cleared the bank by the end of February. The March bank statement had the following information:

March Bank Statement
Steve and Chris Farmer

Beginning cash balance		$ 19,900
Deposits:		
March 1	$120,000	
March 1	50,000	
Total deposits		170,000
Other credits:		0
Checks:		
No. 1000	$ 2,160	
No. 1001	120,000	
No. 1002	46,000	
Total checks		(168,160)
Other debits:		
Bank service charge	$ 15	
Total other debits		(15)
Ending cash balance		$ 21,725

You might notice right away that the beginning balance on the bank statement does not agree with any of the balances in the Farmers' cash account. The check for the tax payment did not clear the bank by the end of February, so as far as the bank is concerned, the cash account appeared as follows at the beginning of March:

1000 Cash

Date	Description	JE#	Debits	Credits	Balance
Beginning Balance					0
Jan. 2	Set up farm bank account	(1)	√20,000		20,000
Feb. 12	Transfer from personal checking account	(3)	√100		19,900

Note that the tax payment cleared the bank in March.

Exercise 7-1 *Begin the reconciliation process for the Farmers and check off the items that appear in both the bank statement and the cash account.*

Answer: When you have finished the comparison, the cash account and the bank statement should appear as follows:

	1000 CASH				
Date	**Description**	**JE#**	**Debits**	**Credits**	**Balance**
Beginning Balance					0
Jan. 2	Set up farm bank account	(1)	√20,000		20,000
Feb. 1	Income tax payment	(31)		√2,160	17,840
Feb. 12	Transfer from personal checking account	(3)	√100		17,940
Mar. 1	Money borrowed to purchase land	(6)	√120,000		137,940
Mar. 1	Money borrowed to purchase tractor	(7)	√50,000		187,940
Mar. 1	Purchase of land	(9)		√120,000	67,940
Mar. 1	Purchase of tractor	(11)		√46,000	21,940
Mar. 10	Sale of bull to neighbor	(10)	900		22,840

Bank Statement
Steve and Chris Farmer

Beginning cash balance		$20,100
Deposits:		
March 1	$120,000√	
March 1	50,000√	
Total deposits		170,000
Other credits:		0
Checks:		
No. 1000	$ 2,160√	
No. 1001	120,000√	
No. 1002	46,000√	
Total checks		(168,160)
Other debits:		
Bank service charge	$ 15	
Total other debits		(15)
Ending cash balance		$21,925

Next, the Farmers classify the unchecked items. The cash account has a deposit that has not cleared the bank. The Farmers sold a bull to a neighbor, received a check, and recorded it on March 10, but did not deposit the check until almost the end of the month. The deposit is a deposit in transit because it has not cleared the bank by the end of March. The Farmers recognize from their bank statement that the bank charged them $15 for bank service charges performed by the bank on their checking account. Now they can prepare the bank reconciliation.

Exercise 7-2 *a. Complete the bank reconciliation for the Farmers. Begin the first section with the ending balance per bank statement. Then list everything in the cash account that did not clear the bank.*

a. Answer:

Bank Reconciliation
For the Period Ending March 31, 20X1

Balance per bank statement		$21,925
Deposits in transit:		
March 10	$900	900
Outstanding checks:		0
Bank errors:		0
Corrected Cash Amount		$22,825

b. Next, begin the second section of the bank reconciliation with the balance in the cash account from the ledger. Then list everything in the bank statement not recorded in the cash account.

b. Answer:

Balance per books		$22,840
Bank charges:		
(Date)	$15	
Total		(15)
Bank credits:		0
Book errors:		0
Corrected Cash Balance		$22,825

The corrected cash balance is the same in each section, so the bank reconciliation is almost complete.

Exercise 7-3 *The final step in reconciliation is to record the items from the second section in journal entries. What journal entry do the Farmers need to record from the March bank reconciliation?*

Answer:

(Date)	6770 Bank Charges	15	
	1000 Cash		15

After posting the journal entry, the cash account shows the corrected cash balance.

1000 CASH					
Date	**Description**	**JE#**	**Debits**	**Credits**	**Balance**
Beginning Balance					0
Jan. 2	Set up farm bank account	(1)	√20,000		20,000
Feb. 1	Income tax payment	(31)		√2,160	17,840
Feb. 12	Transfer from personal checking account	(3)	√100		17,940
Mar. 1	Money borrowed to purchase land	(6)	√120,000		137,940
Mar. 1	Money borrowed to purchase tractor	(7)	√50,000		187,940
Mar. 1	Purchase of land	(9)		√120,000	67,940
Mar. 1	Purchase of tractor	(11)		√46,000	21,940
Mar. 10	Sale of bull to neighbor	(10)	900		22,840
Mar. 31	Bank services charges	(xx)		15	22,825

PRACTICE WHAT YOU HAVE LEARNED   *To practice what you have learned, complete the bank reconciliation exercise in Problem 7-2 at end of the chapter.*

ACCOUNTS RECEIVABLE

As you learned in Chapter 3, the farm accountant records the sale of farm products when receiving the cash from the sale. In Chapter 4, you learned that the year-end adjustments include a journal entry for accounts receivable for money owed to the farm business.

Learning Objective 3 ■ To define and calculate "bad debt expense."

(Date)	Accounts Receivable	XXX	
	Change in Accounts Receivable		XXX

When someone owes money to the farm business, the possibility exists that the party might not pay. The term for these is **bad debts** or uncollectible accounts, but they are not debts of the farm business; rather, they are debts of the parties that owe money to the farm business. Any business with accounts receivable may not receive all of the money for a variety of reasons. Usually, a business is not paid because the owing party is in financial difficulty. If such events occur for a farm business, the financial statements should reflect the loss of income from not receiving the cash from revenue earned.

In a procedure known as the **direct write-off method**, the farm accountant adjusts accounts receivable only when certain that the money will not be received (see Figure 7-4). Certainty exists when the farm business receives notification that the owing party cannot pay or when bankruptcy occurs. When the farm accountant realizes that the cash will not be received, the adjusting journal entry for Accounts Receivable will be for a lower amount than otherwise would have been recorded. The farm accountant normally records revenue from sales on account as part of Change in Accounts Receivable, but when the cash is not forthcoming, the amount of the lost sale is excluded from Change in Accounts Receivable in the year that notification occurs. The income statement reflects the loss of income because it reports the expense of producing the farm products, but no revenue. In the direct write-off method, the adjustment does not occur until after the notification actually occurs. Until that occurs, the Accounts Receivable on the balance sheet reports all of the money owed to the farm business.

Steps involved in the direct write-off method are:

1. Report full amount of accounts receivable on balance sheet.
2. Adjust accounts receivable balance when notification occurs.

According to GAAP and the FFSC Guidelines, the balance sheet should report accounts receivable adjusted for an estimate of lost income from bad debts before notification occurs (see Figure 7-5). This is known as the **allowance method**. If the amount of lost income every year is significant, the farm accountant uses this

FIGURE 7-4 ■ *Direct Write-Off Method for Bad Debts.*

Events:	Sale occurs ⟶ Financial statements prepared ⟶ Notification of bad debt
Accounting:	Accounts Receivable = full amount owed Net Accounts Receivable = amount expected to receive

FIGURE 7-5 ■ *Allowance Method for Bad Debts.*

Events:	Sale occurs ——→	Financial statements prepared ——→	Notification of bad debt
Accounting:		Net Accounts Receivable = amount expected to receive	Net Accounts Receivable = amount expected to receive

method to estimate in advance the amount of bad debts anticipated for the coming year. Rather than wait until the notification occurs, the farm accountant determines the amount of money owed to the farm business and then adjusts it for the amount expected to not be received. In the allowance method, the balance sheet reports the amount that is expected to be received, not the full amount of money owed. By reporting the amount expected to be received, the balance sheet reports a more accurate picture of the assets of the farm business. Without the adjustment, the balance sheet might be misleading. The amount reported on the balance sheet for Accounts Receivable would be "overstated." If the farm accountant anticipates that some of the money owed will not be received and can estimate that amount, the adjustment should be made.

Steps involved in the allowance method are:

1. Calculate the amount of average percentage of bad debts in past years.
2. Calculate the bad debt adjustment for the current year.
3. Calculate net accounts receivable and report it on the balance sheet.

The farm accountant calculates the approximate amount of bad debts by estimating a percentage. The percentage is a fraction of total sales or a fraction of accounts receivable at the end of the year that might not be received. Experience determines the amount of the percentage. Estimating the percentage is difficult if the farm business has not experienced bad debts very often. However, if bad debts are a normal, recurring event, the farm accountant calculates the average percentage of bad debts in the past and applies it to the current year's sales or accounts receivable. The farm accountant has to decide whether to calculate a percentage of sales or a percentage of accounts receivable based on which number is the most consistent from year to year.

Past years: Average percentage of bad debts =
  average amount of lost income from bad debts in past years
  ÷ average sales in past years
Current year: Bad debt adjustment = average percentage of bad debts
  × sales this year

Or

Past years: Average percentage of bad debts =
  average amount of lost income from bad debts in past years
  ÷ average accounts receivable in past years
Current year: Bad debt adjustment = average percentage of bad debts
  × accounts receivable this year

After calculating the bad debt adjustment, the farm accountant subtracts it from the amount owed to the farm business and records the adjusting journal entry or makes an adjustment to cash-basis financial statements.

Net accounts receivable = Accounts Receivable − Bad debt adjustment

Suppose that the Farmers determined that accounts receivable at the end of 20X3 were $12,500. Although they have not experienced any bad debts in the past, Steve and Chris decide to be conservative and adjust their accounts receivable for bad debts. They decide to use an estimate of 1 percent of the accounts receivable amount as the percentage of bad debts. They calculate their bad debt adjustment as follows:

Accounts Receivable	$12,500
Percentage of bad debts	× 1%
Bad debt adjustment	$ 125

The net Accounts Receivable is $12,375:

Accounts Receivable	$12,500
Bad debt adjustment	− 125
Net accounts receivable	$12,375

At the beginning of 20X3, the balance in the Accounts Receivable account was $5,000. Without the bad debt adjustment, the Farmers would calculate the Change in Accounts Receivable as follows:

Amount owed to the Farmers at the end of 20X3	$12,500
Accounts Receivable at the beginning of the year	(5,000)
Change in Accounts Receivable	$ 7,500

Without the bad debt adjustment, the Farmers would record the following journal entry:

Dec. 31 1100 Accounts Receivable	7,500	
4700 Change in Accounts Receivable		7,500

However, with the bad debt adjustment, the amount for the Change in Accounts Receivable is calculated as follows:

Amount owed to the Farmers at the end of 20X3	$12,375
Accounts Receivable at the beginning of the year	(5,000)
Change in Accounts Receivable	$ 7,375

With the bad debt adjustment, the Farmers would record the following journal entry.

(42b)	Dec. 31 1100 Accounts Receivable	7,375	
	4700 Change in Accounts Receivable		7,375

Income Statement:

Cash Sales of Market Livestock and Poultry	$ XXX
+ Change in Accounts Receivable	7,375
Effect on Gross Revenue	$7,375

Ledger:

1100 Accounts Receivable

Date	Description	JE#	Debits	Credits	Balance
Beginning Balance					0
Dec. 31	Adjusting entry to record change in accounts receivable	(42)	3,500		3,500
Dec. 31	Adjusting entry to record change in accounts receivable	(42a)	1,500		5,000
Dec. 31	Adjusting entry to record change in accounts receivable	(42b)	7,375		12,375

Balance Sheet:
    Assets:

Accounts Receivable (net of bad debt adjustment)	$7,375

The notes to the financial statements should include a table of the amounts owed and the due dates of the receivables. If the receivables are commodity receivables, the notes should also include specific information concerning the type of commodity, quantity, and price.

> **PRACTICE WHAT YOU HAVE LEARNED**   *Practice what you have learned and complete Problem 7-3 at the end of the chapter.*

INVENTORY

Learning Objective 4 ■ To explain valuation methods for different categories of inventory.

You learned in Chapters 4 and 5 about the adjustments for different kinds of inventory (feed, crops, and market livestock). You learned that the procedures start with counting the amount of inventory on hand and determining the dollar value. This section discusses the recommendations for estimating the dollar value. You learned that feed, crops, and market livestock can be classified as purchased or raised and that the feed can be classified further as intended for use in the farm operation or to be sold (see Figure 7-6).

Valuation of inventory requires a rearrangement of this classification. Valuation procedures classify the inventory as raised or purchased, and then, for use or sale (see Figure 7-7).

According to GAAP, inventories should be valued at cost except when the current market value is lower than cost. In that case, the inventory should be valued at the market value. According to GAAP, "cost" is the amount spent to produce the inventory and "market" is the cost that it would take to reproduce or replace the item at the current time (the date that the financial statements are prepared). As you read in Chapter 5, the market values for inventories might have to be determined through some appropriate appraisal process.

FIGURE 7-6 ■ *Classes of Inventory.*

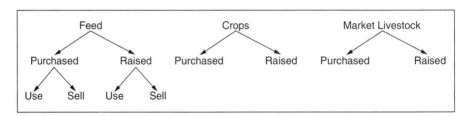

FIGURE 7-7 ■ *Inventory Categories for Valuation.*

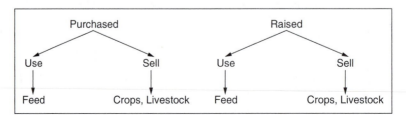

The lower of cost or market valuation comes from a concept of conservatism, which states that when in doubt, the value that is least likely to overstate assets and income should be reported. However, GAAP and the FFSC Guidelines also state that raised crops and market livestock could be valued at **net realizable value** (NRV) when certain conditions exist.

NRV is the estimated sale price (the amount that the inventory could be sold for) less estimated costs of disposal. Costs of disposal refers to any costs of selling the product, such as the cost of delivering the inventory to the marketplace and commission fees.

> The existing conditions for NRV valuation include a reliable, readily determinable, and realizable market price; relatively insignificant and predictable costs of disposal; and, inventory available for immediate delivery.

Inventories Raised for Sale

This category includes crops, market livestock (not livestock to be used for breeding purposes), and feed not intended to be fed to livestock. Because these inventories fulfill the conditions for NRV valuation, NRV is an acceptable value. The actual cost of raising inventory is difficult to determine, but if the farm accountant can do so, the balance sheet would report the cost. A second balance sheet showing NRV values for these inventories can also be prepared. If the cost of raising these inventories cannot be determined, then the balance sheet would report NRV values.

Inventories Raised for Use

This category refers to feed intended for feeding animals and not sold. These inventories do not meet the criteria for NRV valuation. Therefore, the actual cost of producing the feed (cost) or the cost to purchase it at the current time (market), whichever is lower, is the preferred value for these items. If the actual cost to produce the feed is unavailable, the balance sheet would report the feed at the current purchase price.

Inventories Purchased for Resale

This category consists of feeder livestock, feed, and crops intended for resale. Feeder livestock are sold after a period of feeding. Producers may sell the feed in its original form or use it in the production of feeder livestock, milk, eggs, wool, or other farm products. Farm operations might purchase grain or vegetables to fulfill contracts.

Two subcategories classify these types of inventories:

- Items purchased with the intent of being sold in essentially the same form (that is, feeder livestock and crops); and
- Items purchased with the intent of being used for producing other items that will be sold (that is, feed), but could be resold in the same form.

The preferred value for the feeder livestock and the crops is the original cost or the market value, whichever is the lower amount. NRV is acceptable when the cost or market value is not available. For the feed, only the lower of the cost or the market value should be reported. The feed might not be sold if it was not all used up, so it should not be reported at a value that is based on sale price.

Inventories Purchased for Use

Large amounts of seed, fertilizer, fuel, and other supplies could be recorded as inventory purchases rather than expenses. These items are purchased only for use in production. The FFSC Guidelines recommend that these items be valued only at cost. Valuing these items at cost is consistent with GAAP.

TABLE 7-1 ■ *Inventory Categories and Valuation Methods.*

Inventory Category	Class of Inventory	Valuation Method
Raised/Harvested for Sale	Crops	NRV
Raised/Harvested for Sale	Feeder Livestock	NRV
Raised/Harvested for Use	Feedstuffs	Lower of cost or market*
Purchased for Sale	Crops	Lower of cost or market*
Purchased for Sale	Feeder Livestock	Lower of cost or market*
Purchased for Sale	Feedstuffs	Lower of cost or market
Purchased for Use	Supplies	Cost

* NRV is acceptable, but less preferred.

Table 7-1 summarizes the FFSC recommendations for inventory valuation.

The distinction between raised inventory and purchased inventory reflects the ways that the income statement reports them. The income statement reports NRV adjustments for raised inventory in the revenue section as part of the calculation of gross revenue. The NRV method values inventory on potential revenues from selling the inventory, so it makes sense that these adjustments affect gross revenue. The income statement shows purchased items in the operating expense section because they are expenses and, therefore, should be reported at the purchase price, either the original cost or the current purchase price.

The Farmers have the following inventory items at the end of 20X1.

1. Harvested hay on hand—to be used for feeding livestock, but could be sold
2. Harvested grain on hand—to be sold
3. Purchased feed on hand—to be fed to livestock, but could be sold
4. Purchased feeder pigs on hand—to be sold

1. They would classify the harvested hay as inventory raised for use and would value it at the lower of the cost of growing and harvesting the hay or the amount they would have to pay if they bought the hay today. They do not have the cost of producing the hay, so they will value the hay at the market value they would pay if they bought it. In journal entry (33) in Chapter 4, that amount was $2,300.

2. They would classify the harvested grain as inventory harvested for sale and would value it at NRV. In journal entry (34) in Chapter 4, that amount was $6,000.

3. They would classify the purchased feed on hand as inventory purchased for use. Inventory purchased for use is usually valued only at cost, but if it could be sold, it could be valued at lower of cost or market. The feed on hand cost approximately $230. If they bought the feed at the end of the year, they would have to pay $200 for that amount of feed. Therefore, they will value the purchased feed at $200.

4. They would classify the purchased feeder pigs on hand as purchased inventory for sale and value the inventory at the lower of the original cost or the current market price for pigs of the same weight and class as when purchased. The original cost was $750 and the current market value is $700. Therefore, they will value the purchased livestock at $700.

PRACTICE WHAT YOU HAVE LEARNED   *Practice the classification and valuation of inventory by completing Problem 7-4 at the end of the chapter.*

Difficulties with inventory valuation arise when purchased and raised inventory items are mixed together (co-mingled), which is often the case with crops or feed. Feeder livestock can be identified with ear tags, but some farm managers might decide that the extra work involved in keeping records on individual animals is not worth the greater degree of accuracy. When grain is physically co-mingled, the identification of specific bushels by purchase price is virtually impossible because the bushels cannot be separated and identified.

The first issue to consider is the purchased inventory. If inventory was purchased more than once during the year, the possibility exists that different prices were paid for the different purchases. According to GAAP, when this situation occurs, the farm accountant can value the inventory on hand by making assumptions about which ones were sold (and what they cost) and which ones are still on hand (and what they cost). One of the assumptions is **Last-In-First-Out** (LIFO), in which the most recently occurring purchase price is the one used to value the sold items. The inventory that remains is valued according to the oldest purchase price. **First-In-First-Out** (FIFO) assumes that the items sold are part of the first batch purchased; therefore, remaining inventory is valued according to the most recent purchase price.

The FFSC Guidelines do not refer to either of these assumptions, but do mention the use of an average cost of the purchased inventory. The **weighted average method** computes an average cost, taking into account the quantity purchased. The weighted average cost is computed by multiplying the number of units by the price per unit for each of the purchases, then adding the dollar amounts of the purchases and dividing by the total number of units.

The Farmers purchased feeder calves twice during the fall of 20X3. The first purchase, on September 24, consisted of 35 calves weighing a total of 13,500 pounds with a purchase price of $0.65 per pound, or approximately $250 per head. On October 10, the price of feeder calves dropped to $0.60 per pound. After determining that they had sufficient feed on hand to feed more cattle, they purchased an additional 50 head weighing a total of 22,500 pounds for $270 per head. At the end of the year, they still had all the feeder cattle on hand. To determine the value of their feeder cattle inventory, they need to compute the weighted average cost.

$$35 \text{ head} \times 250 = \$\ 8,750$$
$$\underline{50 \text{ head}} \times 270 = \underline{\ 13,500}$$
$$85 \qquad\qquad \$22,250$$

$22,250 \div 85$ head = $262 per head. Therefore, the weighted average cost is $262 per head.

When co-mingling of purchased and raised inventory occurs, the FFSC suggests that the farm accountant assume that all items on hand at the end of the year were purchased. If the number of items purchased is more than the number of items on hand at the end of the year, then all of the items should be valued at either the weighted average purchase cost or the current market value.

If number purchased during the year is greater than the number on hand, then use weighted average cost or market.

If the number of items on hand at the end of the year exceeds the total number that was purchased, then only the number of purchased items are valued at the weighted average cost. The remainder are the number raised and should be valued using the appropriate valuation method for raised inventory for sale.

If number on hand > number purchased during the year, then
   number purchased are valued at weighted average cost and
   number on hand minus number purchased are valued at NRV.

If the Farmers had 15 head of raised feeder cattle on hand when they purchased the feeder cattle in September and October, they would have 100 head. Eighty-five are valued at $262 per head (if market values are greater than cost) and the remaining 15 at NRV at the end of the year (if none of the calves had been sold).

Number on hand (100) is greater than number purchased during the year (85). Number purchased are valued at weighted average cost: 85 @ $262 per head. Number on hand minus number purchased = 100 − 85 = 15, valued at NRV.

If, however, they had sold 20 head before the end of the year, then the remaining 80 head would be valued at $262 (if market values are greater than cost).

Number purchased during the year (85) is greater than number on hand (80). Number on hand are valued at weighted average cost: 80 @ $262 per head.

Under this assumption, it is not necessary to keep the records of the purchased animals separate from the raised cattle, which makes it easier to perform the valuation procedures. When valuing inventory for financial statements, the farm accountant should keep three things in mind:

- The notes to the financial statements should disclose the calculation methods and assumptions. The notes should include quantities in terms of pounds, bushels, tons, and so on, unit values, total values, and the basis for valuation (NRV, lower of cost or market, or cost).
- These methods and assumptions should be used consistently from year to year to enable lenders and other parties to evaluate the financial performance and position of the farm business from one year to the next.
- Conservative methods of valuation should be used for valuing inventory items that will be used or sold during the next year.

PRACTICE WHAT YOU HAVE LEARNED   *Practice the valuation of co-mingled inventory by completing Problem 7-5 at the end of the chapter.*

PREPAID EXPENSES

Prepaid expenses are recorded and valued only at cost. Chapter 4 shows you the accounting procedures for prepaid items in adjusting journal entries (37) and (38) for prepaid insurance and cash investment in growing crops. Disclosures in the notes to the financial statements should list all items recorded as prepaid, the amounts paid, and the amounts used up.

Learning Objective 5 ▆ To recall the procedures for recording prepaid expenses.

CHAPTER SUMMARY

The primary valuation methods for balance sheet items are cost, market value, and net realizable value. The FFSC Guidelines allow market value as an acceptable alternative to historical cost for most assets, both current and non-current. This chapter discusses the issues concerning the valuation of current assets. Although valuation of cash is not an issue of "cost" or "market" value, the chapter emphasizes the importance of performing bank reconciliations so that the farm accountant can keep track of the amount of cash available. Accounts receivable are recorded at an amount expected to be received. Inventory values depend on the class of inventory and its intended use (for sale or for use in production). The readers of farm financial statements require information concerning the valuation methods to understand the balance sheet values. The disclosure notes should contain this information. The next chapter continues the discussion on the valuation of assets.

PROBLEMS

7-1 ▪ Match the following terms with the definitions below.

a. Current market value

b. Current market value method

c. Discounted cash flow method

d. Historical cost

e. Historical cost method

f. Net realizable value method

_____ Basis of valuation from a historical event and can be verified if records exist.

_____ Values are sometimes difficult to obtain or verify.

_____ Method of valuation in which the value shown on the balance sheet is the book value.

_____ Method of valuation based on an exchange in which neither the buyer nor seller are forced to engage in the transaction.

_____ Method of valuation that is particularly relevant for raised crops and livestock.

_____ Method of valuation used for certain types of lease arrangements.

7-2 ■ Prepare a bank reconciliation for the Farmers from the April bank statement and cash account below.

Date	Description	JE#	Debits	Credits	Balance
	1000 CASH				
Beginning Balance					0
Jan. 2	Set up farm bank account	(1)	√20,000		20,000
Feb. 1	Income tax payment	(31)		√2,160	17,840
Feb. 12	Transfer from personal checking account	(3)	√100		17,940
Mar. 1	Money borrowed to purchase land	(6)	√120,000		137,940
Mar. 1	Money borrowed to purchase tractor	(7)	√50,000		187,940
Mar. 1	Purchase of land	(9)		√120,000	67,940
Mar. 1	Purchase of tractor	(11)		√46,000	21,940
Mar. 10	Sale of bull to neighbor	(10)	900		22,840
Mar. 31	Bank service charges	(xx)		√15	22,825
Apr. 1	Employee paycheck	(26)		1,018	21,807
Apr. 17	Purchase of cattle tags	(22)		75	21,732

April Bank Statement
Steve and Chris Farmer

Beginning cash balance		$21,925.00
Deposits:		
April 1	$ 900	
Total deposits		900.00
Other credits:		0
Checks:		
No. 1003	$1,018	
Total checks		(1,018.00)
Other debits:		
Electronic payment for payroll taxes	$274.00	
Electronic payment for FUTA	9.60	
Electronic payment for SUTA	27.60	
Total other debits		(311.20)
Ending cash balance		$21,495.80

7-3 ■ Using the following information, calculate the bad debt adjustment and the net accounts receivable. Then record the adjustment in a journal entry for Accounts Receivable.

Suppose that the Farmers determined that accounts receivable at the end of 20X4 were $10,800. Steve and Chris decide to be conservative and adjust their account receivable for bad debts. They decide to use an estimate of .5 percent of the accounts receivable amount as the percentage of bad debts. Accounts receivable at the end of 20X3 was $12,500.

7-4 ■ Classify the following inventory items and designate the valuation method for each: net realizable value (NRV), lower of cost or market (LCM), or cost (C).

_____Raised feeder pigs   _____Purchased seed for use

_____Purchased feed pellets for use   _____Raised corn crop for sale

_____Raised oats for use   _____Raised wheat crop

_____Purchased feeder calves   _____Raised corn crop for use

_____Purchased fuel   _____Raised hay for sale

7-5 ■ Calculate the value of the co-mingled inventory items in each of the examples below.

First Purchase	Second Purchase	Number Raised	Number at Year-End
a. 25 calves @ $225 per head	100 calves @ $235 per head	0	100
b. 25 calves @ $225 per head	100 calves @ $235 per head	100	150
c. 25 calves @ $225 per head	100 calves @ $235 per head	100	100

CHAPTER 8

Asset Valuation, Part 2— Non-Current Assets

Key Terms

Base value method	Full cost absorption	Individual animal approach	Stocks
Bonds	method	Lump-sum purchases	Transfer points
Cash surrender value	Gains and Losses on Sales	Marketable securities	
Development phase of perennial	of Farm Capital Assets	Productive phase of perennial	
crops	Group value approach	crops	

In Chapter 7, you learned about various valuation methods for current and non-current assets, accounting procedures specific to cash, accounts receivable, inventory, and prepaid expenses, and information disclosed in the notes and tables that accompany the financial statements.

This chapter covers the valuation of non-current assets. Specific issues are the cost of the asset and the allocation of the cost. Non-current assets are not likely sold or used up in a single year, so cost allocation occurs each year that the asset is used. The cost is reported on the balance sheet. The allocation of the cost affects the income statement and the balance sheet. This chapter also addresses the unique topic of valuing raised breeding livestock. According to GAAP, raised breeding livestock should be valued at the full cost of producing and raising the animals. These costs are difficult to verify. Furthermore, growing breeding livestock increase in value up to a point as they age. These issues make raised breeding livestock a different kind of asset that is not typical of non-farm businesses. The FFSC has developed an alternative approach for valuing raised breeding livestock, which you will learn about in this chapter.

This chapter discusses the valuation of perennial crops, the base value approach for valuing raised breeding livestock, the valuation of other non-current assets, such as purchased breeding livestock, machinery and equipment, land, buildings and improvements, and investments, how to allocate the cost of non-current assets over partial periods and with revised estimates, how to account for discarding non-current assets, and how to provide additional information in disclosure notes to accompany the financial statements.

CASH INVESTMENT IN GROWING CROPS AND PERENNIAL CROPS

In Chapter 4, you learned about an account that the FFSC calls the Cash Investment in Growing Crops. The farm accountant should use this account to report the expenses paid each year during the productive phase of perennial crops.

Learning Objective 1 ■ To explain how to value perennial crops during the developing phase.

The **productive phase of perennial crops** refers to the years that the perennial plants are producing a crop. The accounting procedures for the productive phase of perennial crops are explained in Chapter 4.

The FFSC recommends that perennial crops should not be valued at market value prior to harvest. By reporting the value of the perennial crop for the amount spent on producing the crop, the crop is valued at "cost" before harvest. If financial statements have to be prepared after the harvest but before selling the crop, the farm accountant uses the valuation procedures for raised crop inventory that you learned about in Chapter 7. The balance sheet can report the crop at net realizable value (NRV). The market value is the appropriate value after the harvest (see Figure 8-1).

Prior to the productive phase, perennial plants go through a development phase. The **development phase of perennial crops** refers to the time from purchasing and planting the plants until the first year that the plants produce a crop. For some perennial plants, the development phase might be a few years. The costs associated with developing perennial crops include the purchase price of the plants, the costs of planting, and other costs necessary to bring the plants into production. The balance sheet should report the costs of a perennial crop during the development phase as a non-current asset. The allocation of these costs occurs over the estimated life of the plants. An appropriate depreciation method forms the basis for the allocation (see Figure 8-1).

Chapter 3 provides an example in which Steve and Chris had invested in an apple orchard in the year 20X1. They recorded the cost of $45,000 in the Perennial Crops account in journal entry (14). Chapter 4 shows how the Farmers depreciated the perennial crop. If Steve and Chris decided to allocate the cost evenly over the estimated 10-year life for the perennial crop, they would include $4,500 in depreciation expense for the year pertaining to the perennial crop:

$$\text{Depreciation expense for the orchard} = \$45,000 \div 10 = \$4,500$$

When the balance sheet reports book values (assets at cost minus accumulated depreciation), the plants and the land that they reside on should be shown separately on the balance sheet. The balance sheet shows the book value of the plants as the original cost minus accumulated depreciation. The cost of the land is not depreciated. If the balance sheet reports market values, the land and plants should be valued together at market value; it is assumed that they will sell together.

FIGURE 8-1 ■ *Accounting during Development and Productive Phases of Perennial Crops.*

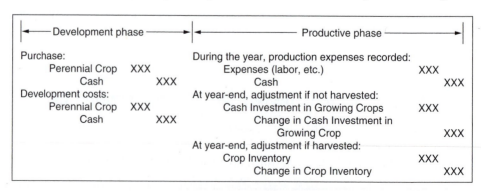

Development phase			Productive phase		
Purchase:			During the year, production expenses recorded:		
Perennial Crop	XXX		Expenses (labor, etc.)	XXX	
Cash		XXX	Cash		XXX
Development costs:			At year-end, adjustment if not harvested:		
Perennial Crop	XXX		Cash Investment in Growing Crops	XXX	
Cash		XXX	Change in Cash Investment in Growing Crop		XXX
			At year-end, adjustment if harvested:		
			Crop Inventory	XXX	
			Change in Crop Inventory		XXX

> PRACTICE WHAT YOU HAVE LEARNED   *Continue the example of the Farmers' perennial crop and perform the accounting procedures for the third year in Problem 8-1 at the end of the chapter.*

RAISED BREEDING LIVESTOCK

The value of some purchased non-current assets is easy to determine if records of the purchases are available. When breeding livestock are raised on the farm, no purchase records exist. The cost values of raised livestock are the expenses of producing these animals. These expenses include feed, labor, livestock tools and supplies, veterinarian services, breeding fees, repairs and maintenance of machinery and fences, fuel, the cost of raising feed, and so on.

The costs of producing breeding livestock are reported as expenses in the years that they occur. The expenses are recorded in expense accounts and nothing is recorded in the breeding livestock account. This result does not reflect the value of the farm's assets because breeding livestock do not have a zero value. The balance sheet should report the cost values for raised breeding livestock. The balance sheet does not report accurate information if not reporting cost values. Furthermore, as you learned in Chapter 4, after raising crops or market livestock, the farm operation reports revenue on the income statement. To be consistent, the farm operation should report revenue from raising the breeding livestock. In addition, unlike other assets that deteriorate or lose value with use, breeding livestock increase in value as they get older, at least for a time, during their productive life. The financial statements should also report this increase in value.

To satisfy the requirement for reporting cost values, the FFSC recommends that farm operations use either the full cost absorption method or the base value method for valuing raised breeding livestock.

> **Learning Objective 2** ■ To define base value accounting for raised breeding livestock and explain the difference between full cost absorption and base value accounting.

- The **full cost absorption** method entails keeping accurate records of all costs associated with the raising of the breeding livestock. The farm accountant keeps track of these costs as the raising occurs.
- The Breeding Livestock account reports the total cost of raising the livestock on the balance sheet, allocated during the useful life of the livestock as depreciation expense, similar to other non-current assets.
- When the livestock sell, the farm accountant determines the amount of gain or loss by comparing the sale price to the book value at the time of the sale.

However, the record keeping required to develop accurate costs of producing the livestock is quite complex. The amount of labor, fuel, and other costs are difficult to allocate to the raising of breeding livestock when these costs also occur for other farm activities.

The FFSC recommends an alternative method for valuing raised breeding livestock.

- The **base value method** designates a base value (obtained from a reputable source) for different categories of raised breeding livestock.
- The farm accountant assigns base values to replacement animals as they enter the breeding herd or flock. The balance sheet reports the total base value of the raised breeding livestock on the balance sheet in the Breeding Livestock account (see Figure 8-2.)
- The animals move through age categories during their normal life cycle. These movements into the next age category are referred to as moving

FIGURE 8-2 ■ *Accounting for Raised Breeding Livestock Using Base Value Approach.*

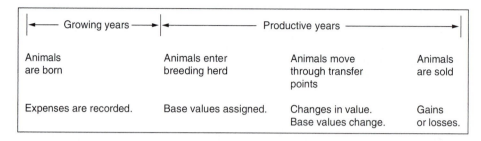

through **transfer points**. A transfer point is the age at which the value of the animal changes.

■ The base value is not allocated as depreciation expense. Instead, changes in value due to age progression occur as animals enter the breeding herd and move through transfer points. The Change in Value Due to Change in Quantity of Raised Breeding Livestock account reports these changes in value, which are included in the calculation of gross revenue on the income statement.

■ When the base values are changed, Gains/Losses Due to Changes in General Base Values of Breeding Livestock reports these changes on the income statement below Net Farm Income from Operations.

■ The Gains/Losses from Sale of Culled Breeding Livestock account reports the gains or losses from sales of raised breeding livestock as part of gross revenue.

The balance sheet can also report market values for these animals. Chapter 5 discusses the accounting for changes in market values.

Figure 8-2 summarizes the following accounting activities for raised breeding livestock:

■ During the growing years, the appropriate expense accounts report the transactions involved in raising the livestock.

■ When animals enter the breeding herd and when they move through transfer points, the farm accountant assigns base values appropriate for the age group.
  • Balance sheet reports total value of breeding livestock in Breeding Livestock account
  • Income statement reports Change in Value Due to Change in Quantity of Raised Breeding Livestock

■ When base values change, the balance sheet and the income statement are affected.
  • Balance sheet reports new total value of breeding livestock in Breeding Livestock account
  • Income statement reports Gains/Losses Due to Changes in General Base Values of Breeding Livestock

■ After selling the animals, the farm accountant calculates the gain or loss.
  • Income statement reports cash received from the sale
  • Gains and losses are reported in Gains/Losses from Sale of Culled Breeding Livestock or in Change in Value Due to Change in Quantity of Raised Breeding Livestock on the income statement

Exercise 8-1 *Review the income statements in Chapter 1 and Appendix A. Where do you find the accounts for raised breeding livestock?*

Answer: The first part of the income statement where Gross Revenue is calculated shows the Gains/Losses from Sale of Culled Breeding Livestock and Change in Value Due to Change in Quantity of Raised Breeding Livestock. Gains/Losses Due to Changes in General Base Values of Breeding Livestock is located further down on the income statement after Net Farm Income from Operations and is used to calculate Accrual Adjusted Net Farm Income.

To begin to use the base value method, the farm accountant must address several issues including the selection of the base values, the animal groupings, individual animals versus groups, distinguishing between the changes in quantity and changes in base values, single versus multiple transfer points, and the long version versus the shortcut alternative for the sale of breeding animals.

Learning Objective 3 ■ To apply the base value approach, using the shortcut method for sales of raised breeding livestock.

Base Values

Base values should represent the approximate cost of raising the animals for their age. The farm accountant can refer to data published by the U.S. Department of Agriculture or land grant institutions to establish the base values for each category. The more sources examined, and the more reliable the sources used, the more likely the base values will approximate the true cost of raising the breeding livestock. When preparing the financial statements, the disclosure notes should indicate the source referred to when assigning base values.

Base values should remain the same for several years. However, the true costs of raising breeding livestock will change over time (most likely increasing because of inflation or other factors). The farm accountant should adjust base values when costs of raising breeding livestock have increased. Changing base values should be based on evidence from the above-mentioned data sources. Frequency of adjustment for base values should be based on the effort exerted to gather the data, and to calculate the gains and losses and the extent of the effect on net income. The income statement reports the change in base values only in the year in which the base values are adjusted (Gains/Losses Due to Changes in General Base Values of Breeding Livestock).

Animal Groupings

The animal groupings must be clearly defined and have some practical and economic basis. Age groupings are an easy and convenient way to form groupings. Another approach is to select the groupings based on categories used to assign market values. An example of groups for a beef cattle herd is the following:

Calves	Under 13 months of age
Yearlings	13 to 24 months
2-year-olds	25 to 36 months
Cows	More than 36 months
Bulls	In service

The farm accountant assigns each category a base value, estimated from reliable data sources.

Individual versus Group Approach

The **individual animal approach** requires maintaining valuation records for each animal. This approach works well for operations that keep production records on individual animals with an identification number for each animal. The disclosure notes should contain summary information on the individual animals. The individual animal approach works in the following way:

- The farm accountant assigns each animal a base value at the time that the animal enters an age group.
- The base value for each individual animal changes only when the animal moves to another category.
- When the base value for a category changes, the animals in that category retain the old base value until they move to the next category. Only new individual animals that enter the changed category are assigned the new base value.

The following example illustrates the assignment of base values using two categories.

Suppose that the Farmers inherited a small herd of raised dairy cattle, consisting of three bred heifers and seven older cows that are in production. They created two groupings: Bred Heifers and Cows. They decided to keep individual valuation records because they planned to keep individual production records. They assigned the Bred Heifer category a value of $800 per head and the Cow category a value of $900 per head. At the end of the first year, they summarized the valuation data in a table as follows:

Individual Animal Approach/Multiple Transfer Points/Long Version/Year 1:

Animal No.	Category	Base Value per Head	Total	New Base Value	Total
98-1	Bred Heifers	$ 800			
98-2	Bred Heifers	800			
98-3	Bred Heifers	800			
96-1	Cows	900			
96-2	Cows	900			
96-3	Cows	900			
96-4	Cows	900			
95-1	Cows	900			
95-2	Cows	900			
94-1	Cows	900			
Total 10 head			$8,700		

As each of the bred heifers freshened the following year, they were moved into the Cow category and assigned a value of $900 each.

The following example demonstrates a change in base values.

Suppose that the Farmers decided to change the base value of the Cow category to $1,000 in the second year. The value of the Bred Heifer category was not changed. If any of the heifers had freshened before the Farmers changed the base value, they would be valued at the old values for the Cow category because they would have moved into the Cow category before the change. Only animals that move into the Cow category after the change are assigned the new base value. Suppose that the heifers had not freshened before they changed the base values. As each heifer freshens, the Farmers assign the new base to that animal. By the end of the year, all of the heifers belong to the Cow category with the new base value. Each of the seven original cows in the Cow category retains the old value of $900 each. They kept two heifer calves from the heifers that freshened and created a new category for Calves (under 1 year of age) with a base value of $400. At the end of the second year, the valuation table appears as follows:

Individual Animal Approach/Multiple Transfer Points/Long Version/Year 2:

Animal No.	Category	Old Base Value	Total	New Base Value	Total
01-1	Calves	$400			$ 400
01-2	Calves	400			400
98-1	Cows	900		$1,000	1,000
98-2	Cows	900		1,000	1,000
98-3	Cows	900		1,000	1,000
96-1	Cows	900			900
96-2	Cows	900			900
96-3	Cows	900			900
96-4	Cows	900			900
95-1	Cows	900			900
95-2	Cows	900			900
94-1	Cows	900			900
Total 12 head			$9,800		$10,100

The difference between the $8,700 value assigned the first year and the old base value of $9,800 at the end of the second year is the result of three bred heifers moving to the Cow category and two new calves entering the herd, verified as follows:

$$\$9,800 - 8,700 = \$1,100$$

Verified: $(900 - 800) \times 3 = \$ \ \ 300$ Change in value of 3 bred heifers moving to Cow category.
$\qquad \$400 \times 2 = \underline{\ \ \ 800}$ Value of 2 new head.
$\qquad\qquad\qquad \$1,100$

They record the following journal entry at the end of the second year for the change in value due to age progression:

Dec. 31   1500 Breeding Livestock   1,100
                  4600 Change in Value due to Change in
                         Quantity of Raised Breeding Livestock   1,100
                         ($9,800 – 8,700 = $1,100)

The difference between the $9,800 old base value and the $10,100 new base value is the change in base values, verified as follows:

$$\$10,100 - 9,800 = \$300$$

Verified: ($1,000 – 900) x 3 = $300 Change in Value of 3 head due to change in base values

They record the following journal entry at the end of the second year for the change in base values:

Dec. 31   1500 Breeding Livestock   300
     8300 Gains Due to Changes in General
       Base Values of Breeding Livestock   300
       ($10,100 − 9,800 = $300)

The ledger account for Breeding Livestock contains the following entries:

1500 Breeding Livestock

Date	Description	Debits	Credits	Balance
Beginning Balance				0
1/2/X1	To set up account for 3 bred heifers and 7 dairy cows	8,700		8,700
12/31/X2	Change in quantity and change due to transfer	1,100		9,800
12/31/X2	Change in base values	300		10,100

The **group value approach** differs from the individual animal approach because individual animals do not have a base value; rather, the number of head in each category is multiplied by the base value for that group. The group value approach works in the following way:

- The farm accountant determines the total base value for a group by multiplying the number of animals in each group by the group's base value at the end of the year.
- The base value for each group of animals changes when the group of animals moves to another category.
- When base value for a category changes, all of the animals in that category have the new base value.

The valuation table for the Farmers' dairy herd at the end of the first year is as follows under the group value approach:

The number of Bred Heifers is three and the number of Cows is seven.

Group Approach/Multiple Transfer Points/Long Version/Year 1:

No. of head	Category	Base Value per head	Total	New Base Value	Total
3	Bred Heifers	$800	$2,400		
7	Cows	900	6,300		
Total 10			$8,700		

At the end of the second year, the number of Calves is two, the number of Bred Heifers is zero, and the number of Cows is 10. All of the Cows have the new base value.

Group Approach/Multiple Transfer Points/Long Version/Year 2:

No. of head	Category	Old Base Value	Total	New Base Value	Total
2	Calves	$400	800		800
10	Cows	900	$9,000	$1,000	$10,000
Total 12			$9,800		$10,800

They record the following journal entries at the end of the second year for the change in value due to age progression:

> Dec. 31   1500 Breeding Livestock   1,100
> 4600 Change in Value Due to Change in
>   Quantity of Raised Breeding Livestock   1,100
>   ($9,800 – 8,700 = $ 1,100)

The journal entry for the change in base values records the difference between the old base values ($9,800) and the new base values ($10,800):

> Dec. 31   1500 Breeding Livestock   1,000
> 8300 Gains Due to Changes in General
>   Base Values of Breeding Livestock   1,000
>   ($10,800 – 9,800 = $1,000)

The ledger account would contain the following balances:

1500 Breeding Livestock

Date	Description	Debits	Credits	Balance
Beginning Balance				0
1/2/X1	To set up account for 3 bred heifers and 7 dairy cows	8,700		8,700
12/31/X2	Change in quantity and change due to transfer	1,100		9,800
12/31/X2	Change in base values	1,000		10,800

The farm accountant can use a shorter way to calculate the amounts for the journal entries for the group approach.

- The farm accountant determines the change in value due to age progression by multiplying the change in the number of head in each age group by the old base values.
- To calculate the change in value due to change in base values, the number of animals in each category affected by the change in base values is multiplied by the change in the base value.

The following example displays a table constructed for the group approach using the shorter method.

In the second year, the base values changed for the cow category. The change in the number of Calves was an increase of 2 head, and 3 Bred Heifers moved to the Cow category, so there are 3 fewer heifers and 3 more cows. The animals affected by the change in base values are the 10 cows, which is the only category with a change in base values. The base value for the Cow category increased by $100. The following table shows the change in value due to age progression and change in value due to change in base values using the shortcut version:

Group Approach/Multiple Transfer Points/Short Version/Year 2/Base Values Changed:

Category	Change in Quantity	Old Base Value	Adjustment	Quantity	Change in Base Value	Adjustment
Calves	2	400	$ 800	2	0	$ 0
Bred Heifers	−3	800	−2,400	0	0	0
Cows	3	900	2,700	10	100	1,000
Total			$1,100			$1,000

The journal entries are the same as those indicated for the group approach using the long version.

Changes in Quantity and Changes in Base Values

As shown in the Farmers' examples above, changes in value of raised breeding livestock can occur as a result of changes in total number of animals (two new calves) or transfers to different age groups (three bred heifers moved to the cow category) and changes in base values (an increase from $900 to $1,000 for the Cow category). Both items appear on the accrual-adjusted income statement. Gross Revenue and Net Farm Income from Operations includes the change in value due to age progression but does not include changes in base values. Changes in base values are included, however, in the calculation of Accrual-Adjusted Net Farm Income. The accrual-adjusted income statement would contain these components for the Farmers' examples using the group value approach:

Change in Value Due to Change in Quantity of Raised Breeding Livestock	1,100
Gross Revenue	$
Operating Expenses +/− Accrual Adjustments	$
Interest Expense +/− Accrual Adjustments	$
Net Farm Income from Operations	$
Gain Due to Changes in General Base Values of Breeding Livestock	1,000
Net Farm Income, Accrual Adjusted	$

Single versus Multiple Transfer Points

The preceding examples illustrate the use of multiple transfer points. Several age categories are set up to organize the values of the raised breeding livestock (Calves, Bred Heifers, and Cows). Changes in value occur as animals move from one age group to another. An alternative approach is the use of a single transfer point. Animals transfer only once when they enter into an "adult" category; that is, when females are bred or when they freshen, and when males enter breeding service. The single transfer point approach occurs as follows:

- Each animal or group has a base value for the adult age group.
- The base value for each individual animal or group does not change due to age progression because the animals do not move into additional categories.

- When base values for the adult category change,
  - In the individual animal approach, only new individual animals that enter the changed category have the new base value.
  - In the group value approach, all of the animals have the new base value.

Prior to adulthood, the animals are immature and valued according to market values. They are not distinguished from other market livestock until they become part of the breeding herd. Only a single category of mature animals has a base value. The single transfer point method is relatively simple to use and requires less book-keeping. The method is useful for herds that do not contain a large number of immature breeding animals relative to the number of other market livestock.

If the Farmers had used a single transfer point, they would categorize the bred heifers as market animals until they freshened. Under the individual animal approach, the following valuation table would appear in the disclosure notes at the end of the first year:

Individual Animal Approach/Single Transfer Point/Long Version/Year 1:

Animal No.	Category	Base Value per Head	Total	New Base Value	Total
96-1	Cows	900			
96-2	Cows	900			
96-3	Cows	900			
96-4	Cows	900			
95-1	Cows	900			
95-2	Cows	900			
94-1	Cows	900			
Total 7 head			$6,300		

At the end of the second year (after the three heifers had freshened), the two new calves would be included with the market animals and the three heifers would now be included in the Cow category:

Individual Animal Approach/Single Transfer Point/Long Version/Year 2:

Animal No.	Category	Old Base Value	Total	New Base Value	Total
98-1	Cows	900		$1,000	$1,000
98-2	Cows	900		1,000	1,000
98-3	Cows	900		1,000	1,000
96-1	Cows	900			900
96-2	Cows	900			900
96-3	Cows	900			900
96-4	Cows	900			900
95-1	Cows	900			900
95-2	Cows	900			900
94-1	Cows	900			900
Total 10 head			$9,000		$9,300

Under the group value approach, the valuation tables would appear as follows for the first and second years, respectively:

Group Approach/Single Transfer Point/Long Version/Year 1:

No. of Head	Category	Base Value per Head	Total	New Base Value	Total
7	Cows	900	$6,300		

Group Approach/Single Transfer Point/Long Version/Year 2:

No. of Head	Category	Old Base Value	Total	New Base Value	Total
10	Cows	900	$9,000	$1,000	$10,000

The values for the bred heifers and calves are included in the account for market livestock until they freshen. If these animals had a low value relative to the value of the rest of the market animals, then including them in this account would have very little effect on the analysis of the financial statements. If, however, the farm operation had very few or no market animals, the analysis could be somewhat distorted.

Shortcut Alternative Version for Animals Sold or Died

The shortcut alternative version refers to valuation when animals sell or have died. The gain or loss from the sale of culled breeding livestock is the difference between the sale price and the base value of the animals sold. The loss from animals that have died is the base value of those animals.

- The base values are determined from the individual animal records under the individual animal approach.
- Under the group value approach, the farm accountant multiples the number of animals sold or died in each category by the base value at the beginning of the year. The farm accountant computes the totals for all categories and compares them with the total amounts received from selling the breeding animals to determine the gain or loss.

The following example of the Farmers' third year with the dairy herd illustrates how the shortcut alternative version works.

In the third year, eight new calves were born and the two calves that were born last year became bred heifers. The Farmers decided to sell one of the bred heifers and two of the cows in the third year (receiving $1,200 for the three head). Using the group approach and multiple transfer points, the valuation table for breeding livestock presents the following information at the end of the third year:

Group Approach/Multiple Transfer Points/Short Version/Year 3/Base Values Unchanged:

No. of Head	Category	Base Value per Head	Total	New Base Value	Total
8	Calves	$ 400	$ 3,200		
1	Bred Heifers	800	800		
8	Cows	1,000	8,000		
Total 17			$12,000		

The change in value due to age progression is $1,200, the difference between the total value at the end of the third year ($12,000) and the total value at the end of the second year ($10,800).

$$\$12,000 - 10,800 = \$1,200$$

Verified:
$1,000 × –2 =	–2000	Sale of two cows
($800-400) × 2 =	800	Transfer of 2 calves to bred heifer category
$800 x –1 =	–800	Sale of 1 bred heifer
$400 × 8 =	3,200	Addition of 8 new calves
	$1,200	

The journal entry for the change in value due to age progression is as follows:

Dec. 31	1500 Breeding Livestock	1,200
	4600 Change in Value Due to Change in Quantity of Raised Breeding Livestock	1,200
	($12,000 – 10,800 = $1,200)	

The income statement would report the following:

Proceeds from Sale of Culled Breeding Livestock	1,200
Change in Value Due to Change in Quantity of Raised Breeding Livestock	1,200
Gross Revenue	$
Operating Expenses +/– Accrual Adjustments	$
Interest Expense +/– Accrual Adjustments	$
Net Farm Income from Operations	$
Gain Due to Changes in General Base Values of Breeding Livestock	$
Net Farm Income, Accrual Adjusted	$

The example above illustrates the shortcut alternative in which the change in value due to age progression is determined by valuing only the change in numbers in each category. A new account (not shown in the chart of accounts in this book), called Proceeds from Sale of Culled Breeding Livestock, reports the amount received from the sale of the culled animals. The gain or loss on the sale is included in calculating the Change in Value Due to Change in Quantity of Raised Breeding Livestock. The advantage of the shortcut alternative is the simplicity of the method. The farm accountant counts only the number in each category on the farm at the end of the year, without having to keep records of the number sold or died. However, gains or losses from culling breeding livestock are not easy to read from the financial statements in the shortcut method.

To keep track of animals died or sold, the farm accountant can use a long version. Appendix I illustrates the long version.

PRACTICE WHAT YOU HAVE LEARNED   *Practice what you have learned and complete Problem 8-2 at the end of the chapter.*

PURCHASED BREEDING LIVESTOCK, MACHINERY AND EQUIPMENT, LAND, BUILDINGS AND IMPROVEMENTS

Learning Objective 4 ■ To identify the expenditures that make up the cost of non-current assets.

The balance sheet shows purchased breeding livestock, machinery and equipment, buildings and improvements, and land at both cost and market values. As you have learned, cost value (or book value) refers to the original cost of the asset minus the total amount of that cost allocated since the purchase of the assets. The allocation is recorded as Depreciation Expense each year and is accumulated in the Accumulated Depreciation account. Accumulated Depreciation is reported on the balance sheet for all of these assets except for land. Land is not depreciated, so the book value is always the original cost. The disclosure notes can display market values (obtained from a qualified appraiser or some other reputable source) in a table, or the farm accountant can prepare two balance sheets (one with the book values and the other with market values).

The original cost of an asset includes the purchase price plus any costs incurred to transport the asset to the farm location and any other costs that occur to get the asset ready to use.

- The original cost of land includes the purchase price, closing costs with the real estate agent, costs of clearing and grading the land, plus the assumption of any liens or mortgages.
- The cost of land does not include improvements with limited lives, such as fences or tiling. These items are classified as improvements and are depreciated.
- The cost of constructing a building includes the materials, labor, and overhead, and any professional fees or permits required. The cost of removing an old building is part of the cost of land, not part of the cost of constructing a new building.
- The cost of machinery and equipment include the purchase price, the freight charges to get the asset to the location on the farm, insurance on the asset while it is being transported, and any installation or assembling costs of the machinery or equipment.

When purchasing a farm, the farm owner purchases the land, buildings, and improvements all together for a single price. These kinds of transactions are called **lump-sum purchases**. Even though a lump sum is paid for all of the assets, the buildings and the improvements must be valued separately from the land to calculate depreciation for each item. The length of the asset's useful life is the basis for depreciation. That basis varies with each type of asset, and with the age of the asset when purchased. In addition, the farm accountant might use different depreciation methods for different assets. The accounting procedures for lump-sum purchases involve the following concepts:

- When purchasing more than one asset together, the farm accountant has to keep track of each asset separately and record an amount for each asset.
- Part of the process of a lump-sum purchase would be an appraisal of each of the assets. If the total price paid (the purchase price) is equal to the sum of the appraised values (market values) of the individual assets, the amount recorded for each asset is its market value.
- If the purchase price for the assets differs from the sum of the individual market values, the farm accountant allocates the purchase price among the assets based on each asset's proportion of the total market value.

In Chapter 3, the Farmers had inherited land and buildings. Suppose that they had purchased the land and buildings for $325,000 instead of inheriting them. The market value for the land and buildings are $240,000 and $110,000, respectively. Because the purchase price of $325,000 differs from the sum of the market values ($240,000 + 110,000 = $350,000), the Farmers have to allocate the purchase price to value the assets. They multiply the proportion of the total market value for each asset by the purchase price to determine the amount of the purchase price allocated to each asset.

	Fair Market Value		Allocation
Land	$240,000	240,000 ÷ 350,000 = 69%	69% × 325,000 = $224,250
Buildings	110,000	110,000 ÷ 350,000 = 31%	31% × 325,000 = 100,750
	$350,000		$325,000

The proportion of the purchase price of $325,000 allocated to the land is $224,250 and the amount allocated to the buildings is $100,750 (See Figure 8-3). A table showing the allocation above should be included in the notes to the financial statements. The journal entry to record the purchase would be as follows:

Jan. 2	1800 Land, Buildings, and Improvements	325,000	
	1000 Cash		325,000

The Farmers might decide to report both the market value and cost value:

	Market	Book Value
Land, Buildings, and Improvements	$350,000	$325,000

The cost of $100,750 for the buildings is depreciated (allocated as depreciation expense) every year. The land is not depreciated.

PRACTICE WHAT YOU HAVE LEARNED   *Practice the lump-sum procedures by completing Problem 8-3 at the end of the chapter.*

LEASED ASSETS

Many farm operations lease non-current assets. Leasing farmland and pastureland is a common practice in the agricultural industry. Leased land is sometimes a large part of the assets of a farm operation. Leased assets require special accounting techniques, which the next chapter, on liabilities, discusses. Because of the nature of some farm leases, the farm accountant records a liability at the beginning of the lease, so the topic is more relevant for the chapter on liabilities. Leased assets are recorded only at cost, not market values. Depreciation is recorded as it is for other non-current assets. Chapter 9 discusses these topics in more detail.

FIGURE 8-3 ■ *Allocation of Lump-Sum Purchase of the Farmers' Land and Buildings.*

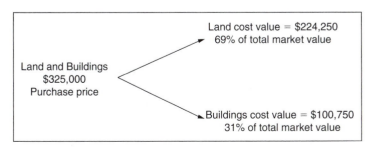

DISPOSING OF NON-CURRENT ASSETS

Learning Objective 5 ■ To perform the accounting procedures for discarding a non-current asset.

The disposing of non-current assets can occur by selling them for cash, exchanging them for other assets, or scrapping them. You learned about the accounting for selling and trading assets in Chapter 3 in journal entries (10) and (11). In those examples, the accounting involves comparing the amount of cash or the market value of the other asset received with the book value of the asset sold or exchanged. The farm accountant records a gain if the value of what is received is greater than the book value of the asset disposed of (for example, the gain on the sale of the Farmers' bull in journal entry (10)). A loss is recorded if the value of what is received is less than the book value of the asset disposed of (for example, the loss on the trade-in of the truck in journal entry (11)). You learned how to calculate gains and losses in Chapter 6 (see Table 6-4). The following is what you have learned so far:

- Gains and losses on the sale of purchased breeding livestock are recorded in the account Gains and Losses of Culled Breeding Livestock, and are reported on the income statement as part of gross revenue.
- Gains and losses on the sale of raised breeding livestock are included in Change in Value Due to Change in Quantity of Raised Breeding Livestock, also a part of gross revenue.
- Gains and losses from the sale or trade-in of all other assets are recorded in the account Gains and Losses on Sales of Farm Capital Assets and are reported on the income statement below operating expenses.

When a farm operation discards an asset, the farm operation does not sell or exchange it, but scraps it instead. The typical assets discarded are worn-out machinery and equipment, too old to find replacement parts for, damaged beyond repair, or too costly to repair. In these cases, the best option is to purchase another one. When an asset is discarded, no cash or other asset is received. In some cases, the farmer has to pay a fee to get rid of the asset. For example, the demolition and removal of a dilapidated old building or hauling away old machinery might involve hiring someone with the necessary tools and equipment to do the job.

The journal entry to record the discarding of an asset involves the following.

- A credit to the asset account for the original cost of the asset.
- A debit to the Accumulated Depreciation account for the total amount depreciated since the asset was put into use.
  - Many of these assets will be fully depreciated by the time that they are discarded, so the amount for Accumulated Depreciation will be the base. As you recall, the base is the amount to be depreciated throughout the entire life of the asset. When an asset is fully depreciated, it means that the base is completely allocated. The asset's book value is the salvage value when it is fully depreciated.
- If cash is paid to remove the asset, a credit is made to the cash account.
- A Loss on Sale of Farm Capital Assets is debited to balance the journal entry. Even though a sale is not involved, this loss account can be used instead of creating another account. Whether the asset is fully depreciated or not, the amount of the loss is equal to the book value of the asset discarded and any amount paid for discarding it.

Suppose that in 2006, the Farmers for some reason had to scrap the truck that was involved in the examples on depreciation methods. They did not trade it in the year 20X6, but rather kept it and purchased a new truck without a trade-in. The truck was fully depreciated by the year 20X6, so Accumulated Depreciation contains the amount of the base ($12,000), and the truck had a book value of $3,000 (the salvage value). If they did not pay anything to discard it, they will record a loss of $3,000 as follows:

(Date)	8200 Loss on Sale of Farm Capital Assets	3,000	
	1980 Accumulated Depreciation	12,000	
	1600 Machinery and Equipment		15,000

If they paid someone $50 to haul the truck away, the journal entry would be:

(Date)	8200 Loss on Sale of Farm Capital Assets	3,050	
	1980 Accumulated Depreciation	12,000	
	1600 Machinery and Equipment		15,000
	1000 Cash		50

> **PRACTICE WHAT YOU HAVE LEARNED** | *Practice what you have learned by completing Problem 8-4 at the end of the chapter.*

INVESTMENTS

Farm operations can invest in many types of entities. The FFSC Guidelines classifies investments as four different types:

- Investments in farm cooperatives
- Investments in entities other than farm cooperatives
- Life insurance policies
- Retirement accounts

> **Learning Objective 6** ■ To explain the valuation of investments in farm cooperatives, investments in entities other than farm cooperatives, life insurance policies, and retirement accounts.

Table 8-1 lists examples of investments in each category. Sometimes the purpose for making investments is to use excess cash to make money, but farm operations make investments for other reasons.

Investments in Farm Cooperatives

The FFSC Guidelines discuss three types of farm cooperatives—supply/manufacturing cooperatives, the Farm Credit System, and marketing cooperatives.

TABLE 8-1 ■ *Examples of Types of Investments.*

CATEGORIES:	FARM COOPERATIVES	OTHER	POLICIES	RETIREMENT
Investments:	Supply/manufacturing cooperatives	Stocks	Life Insurance	Savings accounts
	Farm Credit System	Bonds		Traditional IRAs
	Marketing cooperatives	Land		Roth IRAs

- Supply/manufacturing cooperatives produce or purchase goods for their members and sell them to the members at competitive prices. These cooperatives may distribute patronage refunds when they make a profit.
- Investments in the Farm Credit System involve purchases of membership to obtain credit from a particular lending institution. The lending institutions (also called "credit associations") include the Production Credit Association (PCA), the Federal Land Bank Associations (FLBA), the Federal Land Credit Associations (FLCA), and Agricultural Credit Associations (ACA).
- Marketing cooperatives provide sales outlets for farm products for their members. Members purchase shares of stock in the cooperative that entitle them to deliver a certain amount of product based on the number of their shares held.

The accounting for investments in cooperatives may involve purchasing a membership and recording income.

- If the farm operation purchases a membership, the farm accountant records the payment as a debit to the investments account and a credit to cash.

  (Date)   1900 Investments in Cooperatives and Other
                Investments   XXX
                1000 Cash   XXX
- Farm cooperatives sometimes allocate their profit to the member without a patronage refund. In that case, the farm accountant of each farm member records the amount of profit allocated to that farm operation as an increase to the investment account.

  (Date)   1900 Investments in Cooperatives and Other
                Investments   XXX
                4800 Miscellaneous Revenue   XXX
- If the farm cooperative pays out a patronage refund, the farm accountant records the amount of cash received as a decrease in the investment account. You can think of the investment like a bank account. When some money is received from the account, the amount in the account is reduced.

  (Date)   1000 Cash   XXX
                1900 Investments in Cooperatives and Other
                Investments   XXX

Investments in Entities Other Than Farm Cooperatives

Accounting for these types of investments involves journal entries for the purchase, for income, for market value changes (for some investments), and for selling the investment. Some investments are held for nearly a lifetime and some are held only for short periods, depending on the need for cash or to take advantage of market price changes.

- Investments in stocks and bonds are usually conducted with the aid of a stockbroker or other financial advisor to keep track of market changes and develop strategies for capitalizing on market values. The purpose for investing in stocks or bonds is to receive income or to sell the investment later for a profit.

Stocks are certificates of ownership in a company sold to the public. **Bonds** are IOUs issued by a company for money that the company has borrowed from the public.

The farm financial statements should report the purchase of and any income from these investments. Purchases are recorded in the investments account, just like investments in farm cooperatives.

(Date)	1950 Personal Assets	XXX	
	1000 Cash		XXX

The fund will increase in value due to interest or dividends that accumulate from investing the retirement fund in shares of stock or bonds in mutual funds or other entities. The fund will also increase or decrease as market values of the stocks and bonds change. The purchaser of the retirement fund will receive periodic reports on the changes in value of the retirement fund. An increase in value is recorded in the following way:

(Date)	1950 Personal Assets	XXX	
	3210 Change in Personal Assets		XXX

If losses in the retirement fund occur (because of declining stock market values), the journal entry would be reversed with a credit to the personal assets account and a debit to the change in personal assets.

The FFSC Guidelines recommends that both cost and market values be reported on the balance sheet for retirement accounts. Cost value is equal to the original investment and any additional payments made into the fund. Market value is the cost value plus or minus changes due to fluctuations in the market value of the fund. For all other personal assets, only market values need to be reported.

PRACTICE WHAT YOU HAVE LEARNED   *Practice your knowledge of different types of investments by completing Problem 8-5 at the end of the chapter.*

DISCLOSURE NOTES

The disclosure notes should provide tables of each category of non-current assets containing information on the original cost, accumulated depreciation, book value, and market value for each asset in each category. Each piece of machinery and equipment, each building, each improvement, each parcel of land, each vehicle, each class of raised breeding livestock and purchased breeding livestock, each investment, and each perennial crop should be listed. Each personal asset should be listed, if combined financial statements are prepared. Additional information that is useful for lenders and other parties who read the financial statements include new purchases and sales of assets in each category, with details of the dollar amounts and number of head or number of acres. Table 8-2 summarizes the typical data disclosed in the tables for each class of non-current assets. Lenders often request this kind of information.

Additional information that should be presented in the notes includes:

- The source or method of determining market values.
- The source or method of determining base values for raised breeding livestock.
- Details of the depreciation method used for each class of asset and what the computations were based on—estimated useful life, or total hours, or total miles, and estimated salvage value.
- Reasons for changing estimates or changing from one depreciation method to another.

Learning Objective 7 ▪ To list the information contained in disclosure notes for non-current assets.

TABLE 8-2 ■ *Information Presented in Disclosure Notes.*

BREEDING LIVESTOCK

For Each Age Group:

number of head

number transferred in and transferred out

number sold

number died

base value per head and total base value for raised animals

cost and accumulated depreciation for purchased animals

market value per head and total market value

For the Herd as a Whole:

number of head

total base value for raised animals

total cost and accumulated depreciation for purchased animals

total market value, gains and losses on the sale of culled livestock

calculations for changes in value due to age progression

calculations for changes in value due to changes in base values.

MACHINERY AND EQUIPMENT

For Each of the Following Vehicles, Machinery, and Equipment:

original cost

accumulated depreciation

book value

market value

changes due to additions and disposal

PERENNIAL CROPS

For Each Type of Perennial Crop:

number of acres

original cost and accumulated depreciation

market value

For Perennial Crops as a Whole:

total number of acres

total original cost and accumulated depreciation

total market value

changes due to additions and disposals

LAND, BUILDINGS AND IMPROVEMENTS

For Each Parcel of Land:

year acquired

number of acres

original cost

market value per acre

total market value

TABLE 8-2 ■ (*Continued*)

LAND, BUILDINGS AND IMPROVEMENTS (*Continued*)

For Land as a Whole:

> total number of acres
>
> total original cost
>
> total market value

For Each Type of Building and Improvement:

> year built or acquired
>
> original cost
>
> accumulated depreciation
>
> book value
>
> market value

For Buildings and Improvements as a Whole:

> total original cost
>
> total accumulated depreciation
>
> total book value
>
> total market value
>
> changes due to additions and disposals

INVESTMENTS IN COOPERATIVES AND OTHER INVESTMENTS

For Each Type of Investment:

> book value
>
> market value

For Investments as a Whole:

> total book value
>
> total market value

PERSONAL ASSETS

For Each Type of Personal Asset:

> market values

CHAPTER SUMMARY

Non-current assets can be reported at both cost and market values on the balance sheet. Base values are determined for raised breeding livestock using a variety of approaches concerning the number of transfer points, the choice between the long version and the short version for sales of animals, and the individual animal approach and the group approach. Depreciation expense is reported for all purchased assets using a number of acceptable depreciation methods. For all non-current assets, extensive disclosure information facilitates the analysis of the financial position of the farm business.

PROBLEMS

8-1 ■ Using the information in the chapter for the Farmers' second productive year for the apple orchard and the information below for the third year, determine the amount to be reported on the balance sheet for Cash Investment in Growing Crops. Then calculate the Change in Cash Investment in Growing Crops. Assume that the crop was not harvested by the end of the year and prepare the journal entry. Then prepare a partial income statement and balance sheet for the effects of the third year.

The financial statements for the second year reported the following:

Income Statement:		Cost and Market
Cash Crop Sales	$2,900	
Effect on Gross Revenue		$2,900
Operating Expenses		
+ Change in Investment in Growing Crops	250	
Net Effect on Operating Expenses		250
Net Effect on Net Farm Income from Operations		$3,150

Balance Sheet:		
Assets:		
Cash Investment in Growing Crops		$2,750

The information for the third year includes the following:

Cash sales of second year crop = $3,000
Cash expenditures for third year's crop = $2,600

8-2 ■ Continuing with the Farmers' dairy herd in this chapter and using the information below, calculate Change in Value Due to Change in Quantity of Raised Breeding Livestock and Gains/Losses Due to Changes in General Base Values of Breeding Livestock for Year 4. Start by preparing a valuation table using the group approach, multiple transfer points, and the short version for animals sold.

In the fourth year, seven new calves were born and three of the calves that were born last year became bred heifers. The bred heifer from last year moved into the Cow category. The Farmers sold one of the cows in the fourth year for $450. The base values in each group increased by $50 per head.

8-3 ■ Suppose that the Farmers had purchased the farm for $500,000, which included land, buildings, and improvements. The market value was $250,000 for the land, $100,000 for the buildings, and $50,000 for the improvements. Determine the amount to allocate to the land, buildings, and improvements.

8-4 ■ Suppose that the Farmers decided to tear down all of the buildings that they purchased in Problem 8-3 two years after they purchased the farm because they planned to sell all of the beef and dairy cattle and wanted to set up a hog operation instead. Prepare the journal entry to record the loss on the discarding of the buildings two years after the purchase.

8-5 ■ Identify the type of investment described by the phrases below. Some of the phrases may describe more than one category of investments.

a. investments in farm cooperatives

b. investments in entities other than farm cooperatives

c. life insurance policies

d. retirement accounts

_____ Investing in savings accounts, traditional IRAs, or Roth IRAs.

_____ Investing in an entity that might pay patronage dividends to its members.

_____ Recording the investment by debiting the Investments account.

_____ Recording the share of profit allocated to members by debiting the Investments account.

_____ Recording the dividends by crediting the Investments account.

_____ Investment with a cash surrender value.

_____ Recording changes in market value.

_____ Investing for the sole purpose of earning income or selling the investment for a profit.

_____ Recording the income from the investment by crediting Miscellaneous Revenue.

_____ Investing in marketable securities.

_____ Investing for the purpose of protecting against the loss of key people in the business.

_____ Investing in an entity that produces or purchases goods for its members and sells them to the members at competitive prices.

_____ Purchasing the investment by becoming a member.

_____ Investments that are considered personal assets.

_____ Investments not reported on farm financial statements unless combined statements are prepared.

_____ Investing in an entity that is involved in lending to agricultural operations.

_____ Investing for the purpose of having income in later years.

_____ Investing in an entity that sells stocks and bonds.

_____ Investing in an entity that provides sales outlets for its members.

Valuation of Liabilities and Equity

Key Terms

Amortization schedule	Estimated tax rate	Operating leases	Promissory notes
Bargain purchase option	Installment loans	Present value factor	Tax basis or taxable basis
Capital leases	Lessee	Present value of the lease	Taxable income
Current deferred taxes	Non-Current Deferred Taxes	payments	
Current liabilities	Non-current liabilities		

In Chapter 8, you learned about the valuation of perennial crops, the base value approach for valuing raised breeding livestock, the valuation of other non-current assets, such as purchased breeding livestock, machinery and equipment, land, buildings and improvements, and investments, how to allocate the cost of non-current assets, how to account for discarding non-current assets, and how to provide additional information in disclosure notes to accompany the financial statements.

This chapter discusses liabilities and equity. Liabilities are obligations to outside parties; in other words, the debts of the farm business. A farm business may have obligations to suppliers, lenders, government agencies, and customers. Amounts owed to suppliers include bills for various operating expenses. Amounts owed to lenders include the principal and interest of each loan. Amounts owed to government agencies are various taxes. If a customer pays for something in advance, the farm business owes goods or services to the customer. The farm business also owes money to the owners. We learned in Chapter 1 that this is the equity reported on the balance sheet.

Liabilities are classified into two categories, current liabilities and non-current liabilities. **Current liabilities** are debts due within the next year. **Non-current liabilities** are debts due after the next year. Current liabilities include accounts payable, interest payable, and taxes payable because they are normally due within a short period (less than one year). They also include principal payments on loans due within the next year. Non-current liabilities include the portion of loans not due within the next year (such as mortgage payments). Other liabilities include deferred taxes and capital leases, which have a current portion and may have a non-current portion. The amount of some liabilities is easy to determine. For other liabilities, the farm accountant has to calculate the amount.

In this chapter, you will review the measurement of current liabilities owed to suppliers, government agencies, and lenders that you have learned about in other chapters. You will also learn how to account for installment loans, how to calculate the current and non-current portions of deferred taxes, how to identify and account for capital leases, how to account for owner's withdrawals, valuation equity, and retained capital, and how to disclose in the notes to the financial statements the information relevant for current and non-current liabilities.

ACCOUNTS PAYABLE, TAXES PAYABLE, INTEREST PAYABLE, AND NOTES PAYABLE

Learning Objective 1 ■ To identify common types of liabilities of a farm business.

You have learned about several current liabilities by reading the chapters thus far in this book. From your reading, you know that

- Accounts Payable refers to the amount owed to suppliers for operating expenses. A cost has occurred because the item or service has been received, but if it has not been paid for by the end of the year, the farm accountant reports the amount as Accounts Payable. Examples of Accounts Payable items include utility bills, telephone bills, charges at the local cooperative for supplies, fuel bills, and veterinary bills, to name a few.
- Costs of the farm business also include various taxes. Taxes Payable refers to the amount of taxes owed to various local, state, and federal agencies for income taxes, payroll taxes, and property taxes. If any taxes have not been paid by the end of the year, the farm accountant reports the amounts owed as Taxes Payable.
- When a farm business borrows money from a lending institution, a liability automatically occurs for the principal amount of the loan. If the loan has not been paid off in full by the end of the year, the farm accountant reports the debt as Notes Payable.
- Lending institutions charge interest on loans, so the farm accountant calculates the amount of interest owed at the end of the year, reported as Interest Payable on the balance sheet.

Determining the amount of these liabilities is usually quite straightforward. The amount owed to suppliers is the amount on the purchase receipt or on the bill received from the supplier. Government agencies determine the amount of the taxes and send a bill, or a tax accountant helps the farm accountant calculate these. The principal amount of loans is arranged when the money is borrowed. You learned in Chapter 6 how to calculate the interest on the loans.

Under the accrual-adjusted system, the farm accountant records accounts payable, taxes payable, and interest payable at the end of the year. You learned about these procedures in Chapters 4 and 5.

- The farm accountant compares the amounts owed at the end of the current year to amounts owed at the end of the last year (the beginning of the current year).
- The farm accountant adds the difference in these amounts to the appropriate payable account if the amount owed at the end of the current year is greater than the amount owed at the end of last year. Adjustments in journal entries are recorded as credits to the appropriate payable account. The income statement reports an adjustment as a "change" item. Adjustments in journal entries are debits in the "change" account

FIGURE 9-1 ■ *Expenses Versus Payables.*

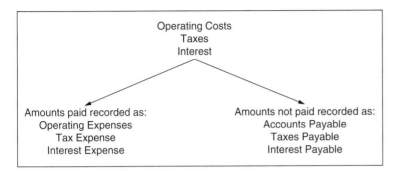

(for example, Change in Accounts Payable). In this case, the "change" items add to the expenses.

■ The reverse is true if the amount owed at the current year is less than the amount owed last year—the payable account is reduced (debited, if journal entries are recorded, and the change account is credited). The "change" item reduces expenses on the income statement.

Chapters 4 and 5 provide illustrations of these adjustments in detail, so they are not repeated here. Figure 9-1 shows the difference between expenses and payables.

The farm accountant should make every effort to keep bills paid up as soon as possible. If bills are not paid when due, some suppliers may charge finance charges, which increase the expenses to the farm business. The percentage rate for these finance charges vary.

You have also learned about the accounting for Notes Payable in Chapters 3 and 6. Unlike Accounts Payable, Taxes Payable, and Interest Payable, which are recorded at the end of the year, the farm accountant records Notes Payable when the money is borrowed. A farm business is likely to have several loans for various purposes; for example, to purchase assets or for operating expenses on a short-term basis. Each of the loans will have different terms for payment period, interest rate, payment amount, collateral, and so on. The farm producer signs promissory notes when the money is borrowed. **Promissory notes** are contracts or IOUs that specify these details and bind the borrower to the promise to pay back the money.

The farm accountant should keep track of each loan separately so that the balance sheet reports the correct amount of liabilities at the end of the year and the income statement reports the correct amount of interest expense for the year. If the entire debt is paid off in one lump sum, the amount of interest paid is the total amount of interest expense on the debt.

> PRACTICE WHAT YOU HAVE LEARNED   *To reinforce these definitions and concepts, complete Problem 9-1 at the end of the chapter.*

Lenders sometimes request periodic payments on loans, which consist of interest and a portion of the principal. These loans are sometimes called **installment loans**. Figure 9-2 illustrates the payment schedule of installment notes compared to lump-sum notes.

Learning Objective 2 ■ To complete the accounting procedures for installment loans.

FIGURE 9-2 ■ *Payment Schedules of Loans (P = Principal, I = Interest).*

Lump-sum payment: P borrowed on March 1, 20X1 and paid back March 1, 20X2
                            20X1   20X2
Jan. Feb. Mar. Apr. May June July Aug. Sept. Oct. Nov. Dec.   Jan. Feb. Mar. Apr. May June

          Mar.1   Mar. 1
           P   P & I
Installment note: P borrowed on March 1, 20X1, payments on the first of each month:
                            20X1   20X2
Jan. Feb. Mar. Apr. May June July Aug. Sept. Oct. Nov. Dec.   Jan. Feb. Mar. Apr. May June

      Mar.1   First of each month
        P   P&I P&I P&I P&I P&I P&I P&I P&I P&I P&I P&I P&I

Installment loans can require one payment per year or several payments per year. The example in Figure 9-2 illustrates payments made on a monthly basis. The farm accountant needs to keep track of the amount of each payment, the interest expense in each payment, and the amount of the principal paid in each payment.

The farm accountant obtains the information about the payments from the loan agreement. The loan agreement specifies

- The principal amount of the loan (the amount borrowed)
- The length of the loan period (the term of the loan)
- The annual interest rate
- The amount of each cash payment
- The total number of payments

Each cash payment is for the same amount. In other words, the farm accountant writes a check for the same amount each time a payment is made.

Farm accountants have to calculate the amount of interest expense in every payment they make. The amount of interest in each payment depends on the unpaid balance of the loan.

Interest expense = Interest rate for the period × Unpaid balance of the loan
Interest rate for the period = Annual interest rate
    ÷ Number of payments in a year
Unpaid balance of the loan = Total principal
    − Amount of principal paid so far

The farm accountant also has to calculate the amount of the principal paid off in each payment. That amount is the difference between the total payment amount and the interest portion of the payment.

Principal portion of payment = Payment − Interest expense

Because a portion of the payment consists of principal, the unpaid balance of the principal is reduced with each payment. Because the unpaid balance is reduced with each payment, the interest portion of each subsequent payment is less than the one before.

Each time a payment is made, the farm accountant records the principal portion of the payment as a reduction of the debt by debiting Notes Payable. In this way, the balance of Notes Payable decreases with every payment. The farm accountant also records the amount of interest paid by debiting Interest Expense every time a payment is made. Lenders should provide an amortization schedule for the appropriate amount for interest expense in each payment. An **amortization schedule** is a table that lists each of the payments and the amount of interest and principal paid with each payment. Alternatively, the farm accountant can calculate the interest and principal portions of each payment with the calculations above.

In Chapter 3, Steve and Chris lease a harvester in 20X1 on a long-term lease contract that allows them to own the harvester at the end of the lease term. Suppose that they purchased the harvester instead for the fair market value of $100,000. Instead of borrowing money from a bank, they can finance the purchase with the dealer. The dealer requires a down payment of $23,982, with annual payments thereafter, and an annual interest rate of 10 percent. The Farmers record the purchase of the harvester in the following way, with a Notes Payable account for the liability:

Year 20X1:

Aug. 1	1600 Machinery and Equipment	100,000	
	2300 Notes Payable—Non-Current		100,000

The down payment is a reduction of the liability:

Year 20X1:

Aug. 1	2300 Notes Payable—Non-Current	23,982	
	1000 Cash		23,982

Subsequent payments for $23,982 include a portion for interest and a reduction of the liability. The Farmers record the next payment as follows:

Year 20X2:

Aug. 1	8100 Interest Expense	7,602	
	2300 Notes Payable—Non-Current	16,380	
	1000 Cash		23,982

The Farmers decide to calculate the amounts for interest and principal and develop an amortization schedule as follows:

Date	Beginning Balance	Cash Payment	Interest	Reduction of Liability	Unpaid Balance
Aug. 1, 20X1	$100,000	$23,982	$ 0	$23,982	$76,018
Aug. 1, 20X2	76,018	23,982	7,602	16,380	59,638
Aug. 1, 20X3	59,638	23,982	5,964	18,018	41,620
Aug. 1, 20X4	41,620	23,982	4,162	19,820	21,800
Aug. 1, 20X5	21,800	23,980	2,180	21,800	0

The beginning balance is the amount of the liability before the payment. Each cash payment is for $23,982. The interest is the beginning balance multiplied by the 10 percent annual rate. The first payment has no interest because it is a down payment on the day they purchased the harvester. They calculate the interest on the second payment as follows:

Unpaid balance of the loan = Total principal − Amount of principal paid so far
= $100,000 − 23,982
= $76,018

Interest rate for the period = Annual interest rate ÷ Number of payments in a year
= 10% ÷ 1
= 10%

Interest expense = Interest rate for the period × Unpaid balance of the loan
$$= 10\% \times 76{,}018$$
$$= \$7{,}602 \text{ (rounded)}$$

They calculate the reduction of the liability in the second payment as follows:

Principal portion of payment = Payment − Interest expense
$$= \$23{,}982 - 7{,}602$$
$$= \$16{,}380$$

They calculate the new unpaid balance after the second payment:

Unpaid balance of the loan = Total principal − Amount of principal paid so far
$$= \$100{,}000 - 23{,}982 - 16{,}380$$
$$= \$59{,}638$$

PRACTICE WHAT YOU HAVE LEARNED   *Complete the procedures for the Farmers' purchase of the harvester by completing Problem 9-2 at the end of the chapter.*

CURRENT DEFERRED TAXES

Learning Objective 3 ■ To determine the amount of current deferred taxes for a farm operation.

In Chapter 4, you read about a current liability called current deferred taxes. You learned that

- The amount of income taxes actually paid by a farm operation depends mainly upon cash-basis income (cash received for farm income and cash payments for operating expenses).
- Income Tax Expense on the income statement is the amount of cash paid for income taxes.

Farm operations usually hire a certified public accountant to determine the amount of income to report on the tax return, called taxable income. **Taxable income**, based on cash-basis income and tax laws, forms the basis for calculating the amount of income tax to pay. Sometimes the farm business has to file its own tax return and sometimes the owners report the income of the farm business on their own tax return. Appendix K explains the general rules for filing tax returns for farm businesses. In this chapter, we assume that the farm business is filing its own tax return and therefore has to report income taxes on the farm financial statements.

- The balance sheet reports the amount of income taxes owed at the end of the year as Taxes Payable. The farm business owes income taxes at the end of the year because the tax returns are not completed and income taxes are not paid until after the end of the year.
- The total amount of taxes on the accrual-adjusted income statement includes Income Tax Expense plus or minus the Change in Taxes Payable.
- Change in Taxes Payable includes the difference between the amount of Taxes Payable at the end of the previous year and the amount of Taxes Payable at the end of the current year. Figure 9-3 illustrates these events.

FIGURE 9-3 ■ *Income Tax Events.*

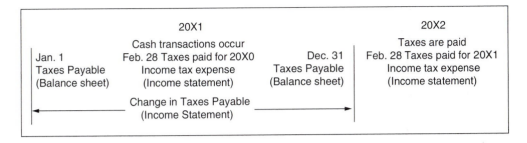

You also learned that

- Accrual Adjusted Net Income on the accrual-adjusted income statement differs from the taxable income prepared for the tax return.
- Reported income tax on the accrual-adjusted income statement should reflect the income tax on accrual-adjusted net income. Not doing so could result in a misstatement of the income taxes of the farm business.

Several items contribute to the differences between cash-basis and accrual-adjusted net income.

- The income statement reports the following cash receipts: crop cash sales, cash sales of market livestock and poultry, livestock products sales, proceeds from government programs, crop insurance proceeds, and cash received from miscellaneous revenue.
- The income statement also includes gains and losses from the sale of culled breeding livestock.
- The accrual-adjusted income statement may also include the following accrual adjustments that affect gross revenue: changes in crop inventories (excluding purchased feed), changes in market livestock and poultry inventories (raised only), and the change in accounts receivable.

When the amount of accrual adjustments is high, the difference between taxable income and accrual-adjusted net income becomes significant.

Similarly, cash-basis expenses differ from the expenses reported on the accrual-adjusted income statement.

- The income statement reports the following cash expenses, which are also tax deductions for the tax return: purchase of feeder livestock, purchase of feed, cash paid for various operating and miscellaneous expenses, cash paid for interest expense.
- The income statement also reports gains and losses on the sale of farm capital assets.
- In addition to these items, the accrual-adjusted income statement may report the following accrual adjustments, which are used to calculate accrual-adjusted net income: change in purchased feeder livestock inventories, change in purchased feed inventories, change in accounts payable, change in prepaid insurance, change in investment of growing crop, change in interest payable, and gains and losses due to changes in base values of raised breeding livestock.

FIGURE 9-4 ▪ *Current Deferred Tax Events.*

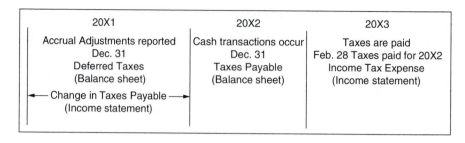

If the amount of the accrual adjustments is significant, they can have a big effect on the difference between taxable income and accrual-adjusted net income.

The accrual adjustments are included in the computation of accrual-adjusted net income, but not considered when calculating taxable income. The tax effects of these items should be included in the reported income tax for accrual-adjusted net income. To exclude them creates an inconsistency and distorts the true amount of income tax. If these items are part of accrual-adjusted net income, then their tax effects should also be part of accrual-adjusted net income.

The balance sheet reports the tax effects of the accrual adjustments as deferred taxes. **Current deferred taxes** are either liabilities or assets related to future tax payments or deductions from the accrual adjustments. These items affect paid income taxes in the future because the cash transactions of these items will occur in the future and will, therefore, either increase or decrease the amount of the tax liability. The income statement reports these tax effects as another part of Change in Taxes Payable (See Figure 9-4).

- These items affect the current year's reported income taxes because they occurred in the current year, but do not affect the amount of paid income taxes until the future event occurs.
- In the following year(s) when the cash transaction occurs, the reported income taxes will not be affected because the income tax related to the item was reported previously; it will, however, affect the amount of paid income taxes.

In Chapter 4, you learned how to make adjustments for current deferred taxes. Table 9-1 repeats those procedures.

TABLE 9-1 ▪ *Procedures for Making Adjustments for Current Deferred Taxes.*

1. Estimate the amount of current deferred taxes for the year.
2. Make adjustment:
   a. Balance sheet shows amount of current deferred taxes

   Current Deferred Taxes
   b. Income statement shows amount of income tax for the year

   (Income Tax Expense)

   +/− Change in Taxes Payable (for difference in Taxes Payable from last year)

   +/− Change in Taxes Payable (for difference in Current Deferred Taxes from last year)

This chapter shows you how to perform the first procedure:

- Estimate the amount of current deferred taxes for the year.

The FFSC Guidelines outline a set of procedures that use balance sheet accounts to estimate deferred taxes. Farm accountants need to understand the definition of tax basis to use these procedures. **Tax basis** or taxable basis, is the amount of cash paid for the asset, minus tax depreciation. Current assets have no depreciation, so for current assets, the tax basis is the balance of the account without accrual adjustments. In the accrual-adjustment system, the tax basis of some current assets is zero because no cash was paid and the current assets do not have a balance until after the accrual adjustments.

The farm accountant also needs to understand the definition of estimated tax rate. The **estimated tax rate** is a percentage multiplied by income to calculate the approximate income tax liability. The tax rate depends on the rates set by the federal and state governments and the level of income, because different levels of income are taxed at different rates. The farm accountant can consult a tax accountant to obtain an estimate of the tax rate for the farm operation. This tax rate applies to the current income tax liability to estimate Taxes Payable and also to deferred taxes to estimate Current Deferred Taxes and Non-Current Deferred Taxes.

Table 9-2 outlines the steps from the Guidelines for calculating and reporting current deferred taxes using balance sheet accounts. Steps 1 through 4 calculate the estimated current deferred taxes. Steps 5 and 6 are the same steps that you learned about in Chapter 4.

TABLE 9-2 ■ *Steps for Calculating Current Deferred Taxes.*

Step 1. The first step in determining current deferred taxes is to determine the difference between the accrual-adjusted values and tax basis values for current assets.

  For each current asset, the tax basis values are subtracted from the accrual-adjusted values.

  The differences between the accrual-adjusted values and the tax basis values are summed.

Step 2. The next step is to determine if there was any income reported on financial statements but not on the tax return for which no asset appears on the balance sheet. According to the FFSC Guidelines, some examples of this income include crop insurance proceeds and disaster payments.

Step 3. The next step is to determine the amount of accrual-adjusted current liabilities, such as accounts payable.

Step 4. Add the results of Steps 1 and 2.

  Subtract the total of Step 3 from the sum of Steps 1 and 2.

  Multiply the result times the estimated tax rate. The result is the amount of Current Deferred Taxes at the end of the year.

Step 5. Subtract the balance in the Current Deferred Taxes account at the beginning of the year from the answer in Step 4.

Step 6. Prepare the journal entry or make the adjustment directly to the income statement with the answer in Step 5 as you learned in Chapter 4, using the Change in Taxes Payable account.

In Chapter 4, the Farmers made adjustments in journal entries (32) to (42) (except for (39)) that resulted in several current asset and current liability accounts reported on the 20X1 balance sheet.

Balance Sheet:
    Assets:

Accounts Receivable	$3,500
Feed Inventory Purchased for Use	230
Feed Inventory Raised for Use	2,300
Crop Inventory Raised for Sale	6,000
Feeder Livestock Inventory Purchased for Resale	750
Feeder Livestock Inventory Raised for Sale	50,000
Prepaid Expenses	700
Cash Investment in Growing Crops	2,500

    Liabilities:

Interest Payable	5,000
Accounts Payable	340

The Farmers did not have any income such as crop insurance proceeds or disaster payments in 20X1. The tax basis for each of the current assets and current liabilities is zero. After consulting with a tax accountant, they decided to use an estimated tax rate of 30 percent. The Farmers applied the six steps for current deferred taxes as follows:

Step 1. They calculated the difference between the accrual-adjusted values and tax basis values for current assets.

Current Assets	Balance Sheet Value	Tax Basis	Difference
Accounts Receivable	$3,500	$0	$ 3,500
Feed Inventory Purchased for Use	230	0	230
Feed Inventory Raised for Use	2,300	0	2,300
Crop Inventory Raised for Sale	6,000	0	6,000
Feeder Livestock Inventory Purchased for Resale	750	0	750
Feeder Livestock Inventory Raised for Sale	50,000	0	50,000
Prepaid Insurance	700	0	700
Cash Investment in Growing Crops	2,500	0	2,500
Total			$65,980

Step 2. They had no crop insurance proceeds or disaster payments.

Step 3. They listed the amount of accrual-adjusted current liabilities.

Current Liabilities	Balance Sheet Value
Interest Payable	$5,000
Accounts Payable	340
Total	$5,340

Step 4. They subtracted the total of Step 3 from the sum of Steps 1 and 2 and multiplied the result by the estimated tax rate of 30 percent.

Steps 1 and 2	$65,980
Step 3	− 5,340
Difference	$60,640
Tax rate	× 30%
Current Deferred Taxes	$18,192

The result is the amount of Current Deferred Taxes at the end of the year.

Step 5. They subtracted the balance in the Current Deferred Taxes account at the beginning of the year from the answer in Step 4.

Current Deferred Taxes at the end of the year	$18,192
− Current Deferred Taxes at the beginning of the year	− 0
Change in Taxes Payable	$18,192

Step 6. They recorded the following journal entry:

Dec. 31	9110 Change in Taxes Payable	18,192	
	2510 Current Deferred Taxes		18,192

Their financial statements reported the following:

Income Statement:

Income before Taxes	$ XXX
Income Tax Expense	(2,160)
+ Change in Taxes Payable	(870)
+ Change in Taxes Payable	(18,192)
Effect on Accrual Adjusted Net Income	($21,222)

Balance Sheet:
Liabilities:

Taxes Payable	$ 3,030
Current Deferred Taxes	18,192

PRACTICE WHAT YOU HAVE LEARNED   *Help the Farmers estimate the current deferred taxes for the following year, 20X2, by completing Problem 9-3 at the end of the chapter.*

NON-CURRENT DEFERRED TAXES

In Chapters 4 and 5, you learned about the nature of non-current deferred taxes.

Learning Objective 4 ▪ To determine the amount of non-current deferred taxes for a farm operation.

- Non-current deferred taxes arise from the differences between the book or base values of non-current assets and their tax basis, and from the differences between the market values of non-current assets and their book or base values.
- The balance sheet reports these tax liabilities as **Non-Current Deferred Taxes** because they are not likely to be resolved during the next year.

First Component of Non-Current Deferred Taxes

A non-current deferred tax may consist of two components. The balance sheet reports the liability for the first component as part of Non-Current Deferred Taxes. The adjustment on the income statement for the first component is part of Change in Taxes Payable. The first component relates to

- the difference between base values and tax basis of raised breeding livestock; and
- the difference between book values and tax basis of purchased non-current assets.

Determining the non-current deferred taxes for raised breeding livestock requires calculating the difference between the base values and the tax basis. As you learned in Chapter 8, when a farm operation raises breeding livestock, the costs of raising the livestock are recorded in expense accounts, not the Breeding Livestock account. Base values are assigned and reported in the Breeding Livestock account. When raised breeding livestock have base values for financial reporting purposes, the tax basis is zero. Tax basis is based on the cash paid for an asset, but no cash purchase occurred for raised livestock, so no tax basis exists. Therefore, the difference between the base value and the tax basis is simply the base value.

When depreciation methods for financial accounting and for tax purposes differ, the book value is different from the tax basis.

- Book Value = Original cost of an asset − Accumulated financial depreciation of the asset. Accumulated financial depreciation is based on depreciation methods for financial accounting.
- Tax Basis = Original cost of an asset − Accumulated tax depreciation of the asset. Accumulated tax depreciation is based on tax depreciation methods.

Depreciation expense on the accrual-adjusted income statement is different from depreciation on the tax return, so accrual-adjusted net income is different from taxable income. This means that income taxes paid on the taxable income do not reflect the income taxes on accrual-adjusted net income. The farm accountant has to perform an adjustment so that the income taxes on accrual-adjusted net income include the effects of the depreciation method used to calculate accrual-adjusted net income. Remember, Income Tax Expense is the amount of taxes paid (based on taxable income). Income Tax Expense along with the adjustments reports the income taxes on accrual-adjusted net income.

In Chapter 4, you learned about adjusting for the first component of non-current deferred taxes. Table 9-3 reviews these procedures.

This chapter shows you how to perform the first procedure.

- Estimate the amount of the first component of non-current deferred taxes for the year.

Table 9-4 outlines the procedures for calculating the first component of non-current deferred taxes. Steps 1 through 3 show how to estimate the amount of the

TABLE 9-3 ■ *Procedures for Making Adjustments for the First Component of Non-Current Deferred Taxes.*

---

1. Estimate the amount of the first component of non-current deferred taxes for the year.

2. Make adjustment:

   a. Balance sheet shows amount of non-current deferred taxes
   Non-Current Deferred Taxes

   b. Income statement shows amount of income tax for the year
   (Income Tax Expense)
   +/− Change in Taxes Payable (for difference in Taxes Payable from last year)
   +/− Change in Taxes Payable (for difference in Current Deferred Taxes from last year)
   +/− Change in Taxes Payable (for difference in first component of Non-Current Deferred Taxes from last year)

TABLE 9-4 ■ *Steps for Calculating First Component of Non-Current Deferred Taxes.*

Step 1. Make adjustments for change in value due to age progression and change in base values (if applicable) for raised breeding livestock (if not already done).

Calculate the difference between the base value and the tax basis for raised breeding livestock.

Step 2. Make adjustment for depreciation on purchased non-current assets (if not already done).

Calculate current book values of purchased non-current assets.

Book value = Original cost – Accumulated depreciation

Obtain tax basis of purchased non-current assets from tax accountant.

Calculate the difference between the book value and the tax basis for purchased non-current assets.

Step 3. Add the results from Steps 1 and 2 and multiply the sum times the estimated tax rate.

The result is the amount of Non-Current Deferred Taxes for the first component.

Step 4. Calculate the difference between the result from Step 3 and the amount of Non-Current Deferred Taxes at the beginning of the year that pertains to the first component.

Step 5. Prepare the journal entry or make the adjustment directly to the income statement with the answer in Step 4 as you learned in Chapter 4, using the Change in Taxes Payable account.

first component. Steps 4 and 5 are the same steps that you learned about in Chapter 4. For the raised breeding livestock, the farm accountant starts with the total base value of the raised breeding livestock and makes adjustments for changes in the value due to age progression and changes in base values (if not already done) to determine the current base value. Then, the farm accountant determines the current book value of purchased breeding livestock and other non-current assets by making the adjustment for depreciation (if not already done). For these purposes, values of raised breeding livestock and purchased breeding livestock should be kept separate. The farm accountant obtains the tax basis from the tax accountant, who calculates tax depreciation for the farm's assets.

Step 1. The example of the Farmers' dairy cattle herd in Chapter 8 illustrates these procedures. Using the individual animal approach with multiple transfer points for Year 1 (assume that Year 1 is 20X1), the Farmers begin by determining the current base value of the dairy herd. Assume that at the beginning of 20X1, they did not report a value for the dairy herd because they inherited the cattle during the first year of operation. Using journal entries, the Farmers record the following journal entry at the end of 20X1 for changes in value due to age progression to establish the base values:

Dec. 31	1500 Breeding Livestock	8,700	
	4600 Change in Value Due to Change in		
	Quantity of Raised Breeding Livestock		8,700
	($8,700 – 0 = $8,700)		

The result of this entry is a current base value of $8,700 for the raised breeding livestock:

1500 Breeding Livestock

Date	Description	Debits	Credits	Balance
Beginning Balance				0
12/31/X1	To set up account for 3 bred heifers and 7 dairy cows	8,700		8,700

They calculate the difference between the base value and the tax basis as follows:

Base value	$8,700
Tax basis	0
Difference between base value and tax basis	$8,700

Step 2. In Chapter 4, journal entry (39), the Farmers recorded depreciation expense at the end of 20X1. Accumulated depreciation is only for one year's depreciation at the end of their first year of operation. Then they calculate the book values.

Asset	Cost	Accumulated Depreciation	Book Value
Office Furniture	$ 1,000	$ 250	$ 750
Buildings, Improvements	110,000	4,400	105,600
Tractor	50,000	3,333	46,667
Truck	15,000	3,000	12,000
Leased Harvester	100,000	6,667	93,333
Orchard	45,000	4,500	40,500
Total		$22,150	$298,850

They obtain the tax basis from the tax accountant and calculate the difference between the book values and the tax basis:

Asset	Book Value	Tax Basis	Difference
Office Furniture	$ 750	$ 600	$ 150
Buildings, Improvements	105,600	100,000	5,600
Tractor	46,667	38,000	8,667
Truck	12,000	9,000	3,000
Leased Harvester	93,333	93,333	0
Orchard	40,500	36,000	4,500
Total	$298,850	$276,933	$21,917

Step 3. They calculate the sum of Steps 1 and 2 and the non-current deferred taxes for the first component using the same estimated tax rate that they used for the current deferred taxes:

Step 1	$ 8,700
Step 2	21,917
Sum	$30,617
Tax Rate	× 30%
Non-Current Deferred Taxes	$ 9,185

They report this amount on the balance sheet along with the amount for the second component of non-current deferred taxes.

Step 4. They calculate the difference between the result from Step 3 and the amount of Non-Current Deferred Taxes at the beginning of the year that pertains to the first component. Because 20X1 is the first year of operation for the Farmers, the beginning balance is zero.

Non-Current Deferred Taxes at the end of the year (first component)	$9,185
− Non-Current Deferred Taxes at the beginning of the year	− 0
Change in Taxes Payable	$9,185

Step 5. They record the following journal entry:

Dec. 31	9110 Change in Taxes Payable	9,185	
	2500 Non-Current Deferred Taxes		9,185

Their financial statements report the following so far:

Income Statement:

Income before Taxes	$ XXX
Income Tax Expense	(2,160)
+ Change in Taxes Payable	(870)
+ Change in Taxes Payable	(18,192)
+ Change in Taxes Payable	(9,185)
Effect on Accrual Adjusted Net Income	($30,407)

Balance Sheet:

Current Liabilities:	
Taxes Payable	$ 3,030
Current Deferred Taxes	18,192
Non-Current Liabilities:	
Non-Current Deferred Taxes	9,185

PRACTICE WHAT YOU HAVE LEARNED   *Practice what you have learned and complete Problem 9-4 at the end of the chapter.*

Second Component of Non-Current Deferred Taxes

In Chapter 5, you learned that the second component of non-current deferred taxes relates to the difference between book or base values and current market values of non-current assets.

- The difference between current market values and base values of raised breeding livestock; and
- The difference between current market values and book values of purchased non-current assets.

The farm accountant reports this component only if the balance sheet reports market values.

The liability for deferred taxes occurs because when assets are sold in the future, they will be sold for market value. As you have learned throughout this book, a gain or loss usually occurs because the market value received is seldom the same as the book value. Gains or losses are not reported for tax purposes because they do not represent cash income or deductions. However, they are a component of accrual-adjusted net income, so their tax effects should also be reported. The taxes are deferred because the sale has not occurred yet. Figure 9-5 shows you the events associated with the second component of non-current deferred taxes.

The farm accountant makes a deferred tax adjustment while preparing market-based financial statements to report the tax effects of market value adjustments. The balance sheet reports the deferred taxes as the second component of Non-Current Deferred Taxes. The difference between the amounts of non-current deferred taxes from the previous year to the current year is not reported on the income statement, but rather is reported on the statement of owner equity. The statement of owner equity reports the change in equity for market valuation (as Change in Excess of Market Value

FIGURE 9-5 ■ *Non-Current Deferred Tax Events (Second Component).*

20X1	Future	Following Year
Market values are reported Dec. 31 Deferred Taxes (Balance sheet) Change in Non-Current Deferred Taxes (Statement of Owner Equity)	Assets are sold Dec. 31 Taxes Payable (Balance sheet) Gain/Loss on Sale (Income Statement)	Taxes are paid Feb.28 Taxes paid for 20X2 Income Tax Expense (Income statement)

over Cost/Tax Basis of Farm Capital Assets), so it should also report the tax effects of these changes. Table 9-5 outlines the procedures that you learned about in Chapter 5. This chapter shows you how to perform the first procedure.

■ Estimate the amount of the second component of non-current deferred taxes for the year.

Table 9-6 outlines the procedures for calculating the second component of non-current deferred taxes. Steps 1 through 3 show you how to estimate the amount of the first component. Steps 4 and 5 are the same steps that you learned about in Chapter 5. After working on the first component, the farm accountant has the book values and base values. An appraisal from a qualified farm appraiser can provide the market values.

TABLE 9-5 ■ *Procedures for Making Adjustments for the Second Component of Non-Current Deferred Taxes.*

---

1. Estimate the amount of the second component of non-current deferred taxes for the year.

2. Make adjustment:

   a. Balance sheet shows amount of non-current deferred taxes
      Non-Current Deferred Taxes

   b. Statement of owner equity shows the market value adjustments and the tax effect of the market value adjustments for the year
      Change in Excess of Market Value over Cost/Tax Basis of Farm Capital Assets
      +/– Change in Non-Current Portion of Deferred Taxes

TABLE 9-6 ■ *Steps for Calculating Second Component of Non-Current Deferred Taxes.*

Step 1. Determine market values for all non-current assets from a farm appraiser.

   Calculate the difference between the market value of each purchased non-current asset and their book value.

Step 2. Calculate the difference between the market value and the base value for raised breeding livestock.

Step 3. Add the results from Steps 1 and 2 and multiply times the estimated tax rate.

   The result is the amount of Non-Current Deferred Taxes for the second component.

Step 4. Calculate the difference between the result from Step 3 and the amount of Non-Current Deferred Taxes at the beginning of the year that pertains to the second component.

Step 5. Prepare the journal entry or make the adjustment directly to the statement of owner equity with the answer in Step 4 as you learned in Chapter 5, using the Change in Non-Current Deferred Taxes account.

Chapter 5 presents an example of market value adjustments for the Farmers' non-current assets. The Farmers calculate book values by subtracting the accumulated depreciation for each asset (except land and raised breeding livestock) from its cost. They also conducted an appraisal of their non-current assets. The land that Steve and Chris inherited had a value of $800 per acre at the time of the inheritance. It consists mostly of cropland. Similar land in the area recently sold for $900 per acre. The recently purchased acreage has not changed much in value since the purchase. The machinery and equipment have a book value of $58,667 and a total market value of approximately $62,167. The 145 breeding cows have a base value of $79,500 (after the adjustment for age progression) and a market value of $64,500. The appraisal indicates that the market values for other assets are as follows: office furniture and equipment, $750; orchard, $40,500; buildings and improvements, $105,600; and leased harvester, $93,333. Then they compute the difference between the book or base values and the market values.

Steps 1 and 2. They calculate the difference between the market value of each non-current asset and its book or base value.

Non-current assets (cost − accumulated depreciation)	Book/Base Values	Market Values	Difference
Breeding Livestock ($73,500 + 6,000 age adjustment)	79,500	64,500	(15,000)
Machinery and Equipment ($65,000 − 3,333 − 3,000)	58,667	62,167	3,500
Office Equipment and Furniture ($1,000 − 250)	750	750	0
Perennial Crops ($45,000 − 4,500)	40,500	40,500	0
Land ($240,000 + 120,000; no depreciation)	360,000	390,000	30,000
Buildings/Improvements ($110,000 − 4,400)	105,600	105,600	0
Leased Assets ($100,000 − 6,667)	93,333	93,333	0
Excess of market value over cost			18,500

Step 3. They multiply the results from Steps 1 and 2 times the estimated tax rate.

Steps 1 and 2	$18,500
Tax rate	× 30%
Non-Current Deferred Taxes	$ 5,550

The result is the amount of Non-Current Deferred Taxes for the second component.

Step 4. They calculate the difference between the result from Step 3 and the amount of Non-Current Deferred Taxes at the beginning of the year that pertains to the second component. Because 20X1 is the first year of operation for the Farmers, the beginning balance is zero.

Non-Current Deferred Taxes at the end of the year (second component)	$5,550
− Non-Current Deferred Taxes at the beginning of the year	− 0
Change in Non-Current Deferred Taxes	$5,550

Step 5. They record the following journal entry:

Dec. 31	3020 Change in Non-Current Deferred Taxes	5,550	
	2500 Non-Current Deferred Taxes		5,550

Their financial statements report the following:

Income Statement:

Income before Taxes	$ XXX
Income Tax Expense	(2,160)
+ Change in Taxes Payable	(870)
+ Change in Taxes Payable	(18,192)
+ Change in Taxes Payable	(9,185)
Effect on Accrual Adjusted Net Income	($30,407)

Statement of Owner Equity:

Valuation Equity:

Change in Excess of Market Value over Cost/Tax Basis of Farm Capital Assets	$18,500
Change in Non-Current Deferred Taxes	(5,550)
Effect on Equity	$12,950

Balance Sheet:

Current Liabilities:

Taxes Payable	$ 3,030
Current Deferred Taxes	18,192

Non-Current Liabilities:

Non-Current Deferred Taxes (9,185 + 5,550)	14,735

Equity:

Valuation Equity	12,950

Both components of the non-current portion of deferred taxes will exist when the book or base values differ from the tax basis and the market values differ from the book values. The examples show that the farm accountant has to keep records of both components separately. If a beginning balance for Non-Current Deferred Taxes exists, the farm accountant needs to know the amount of the beginning balance that pertains to each component.

> **PRACTICE WHAT YOU HAVE LEARNED** | *Practice what you have learned and complete Problem 9-5 at the end of the chapter.*

CAPITAL LEASES

Learning Objective 5 ■ To explain the requirements for reporting a lease as a capital lease.

Farm operations may lease assets rather than purchase them. In some cases, the farm operation leases an asset for a very short period (for example, a machine for a day or a certain number of hours). In some cases, the farm operation leases an asset for many years (for example, land). The accounting for the lease depends on the nature of the lease arrangement. Many lease arrangements qualify as **operating leases**, which are rental arrangements that do not meet the criteria for a capital lease. The farm accountant records the rental payments for an operating lease as rent expense.

Capital leases are special types of leases that, in effect, are installment purchases. With most installment purchases, the buyer obtains ownership (the title) of the asset at the time of the purchase and then makes periodic payments. Many characteristics of a capital lease are very similar to an installment purchase. With capital leases, the **lessee** (the party that is using the asset) makes periodic payments similar to installment payments. The lessee might obtain ownership of the asset. Not all capital leases result in the lessee owning the asset. If the lessee obtains ownership, the transfer of the title will not happen until the end or near the end of the lease term. The arrangement is a lease because during the lease period, the ownership of the asset remains with the original owner.

A capital lease differs from an operating lease because in a capital lease, the lessee has all of the benefits and risks of the owner of the asset.

- In a capital lease, the lessee can report the leased asset along with other owned assets on the balance sheet and can record depreciation expense. In an operating lease, the lessee does not report the asset on the balance sheet and does not record depreciation.
- In a capital lease, the lessee also reports a liability for all of the lease payments in the future. In an operating lease, the lessee does not report a liability.
- In a capital lease, the lessee records lease payments, similar to loan payments. In an operating lease, the lessee only records rental expense for the payments.
- Often in a capital lease, the lessee is responsible for repairs and maintenance of the asset. Generally, in an operating lease, the responsibility for repairs and maintenance remains with the owner of the asset.

A lease is a capital lease when the following conditions are present:

- The lease is a non-cancelable lease contract, that is, the lease cannot be canceled.
- The lease contract specifies that the lessee will make a series of payments for a specific period in return for the use of a specified asset.
- The lease agreement meets one of four ownership tests.

The ownership tests determine whether the lessee is essentially acting as the owner of the asset. The farm lessee examines the lease agreement for ownership tests when the lease begins so that the farm accountant can perform the proper accounting for the lease.

Table 9-7 outlines the four ownership tests. In the first test, if the contract specifically states that the title will pass to the farm lessee at the end of the lease, the farm lessee will obviously become the owner of the asset and the lease arrangement is really a purchase. If the lease agreement does not specify a transfer of title, the farm accountant checks for the second test. In the second test, the farm accountant looks for an opportunity to buy the asset at a bargain price. Even though the lease contract does not specify ownership transfer explicitly, the farm lessee very likely will take advantage of purchasing the asset for a bargain. Again, the farm lessee looks like an owner. If the lease agreement does not pass the first two tests, the farm accountant checks the third test. In the third test, if the farm lessee uses the asset for 75 percent or more

TABLE 9-7 ■ *Ownership Tests for Capital Leases.*

1. The lease agreement specifies that the ownership of the asset transfers to the lessee.

2. The lease agreement contains a bargain purchase option. A **bargain purchase option** offers the lessee the option to purchase the asset at the end of the lease term for a price well below the estimated fair market value of the asset at the end of the lease term.

3. The lease term is equal to 75 percent or more of the expected economic life of the asset. The expected economic life of an asset is subjective and a reasonable estimate should be used. Some lease arrangements may contain a renewal option. In that case, the lease term in question is the initial term plus any amount of time covered by renewal periods.

4. The present value of the lease payments is equal to or greater than 90 percent of the fair market value of the asset. This ownership test will also not apply if the lease begins during the last 25 percent of the asset's economic life.

of its useful life, the farm lessee has had possession for the vast majority of the asset's life, similar to an owner. When the condition for the fourth test occurs, the farm lessee is virtually paying for the asset and, even if the contract does not specify that title passes to the lessee, the series of payments resembles an installment purchase.

> **PRACTICE WHAT YOU HAVE LEARNED** *To help you remember these concepts, complete the matching exercise in Problem 9-6 at the end of the chapter.*

Learning Objective 6 ■ To perform the accounting procedures for a capital lease.

When a lease is classified as a capital lease, the balance sheet of the farm lessee reports a leased asset and a lease liability. The owner and the farm lessee agree on the amount of the payments. The value of the asset is depreciated by the farm lessee.

■ The first step at the start of the lease is to determine the value of the asset and report this amount as Leased Asset and as Obligations on Leased Assets. In most cases, the value to record will be the fair market value of the asset.

| Leased Asset | XXXX | |
| Obligations on Leased Asset | | XXXX |

■ The farm lessee makes and records the payments every month or every year, according to the lease arrangement. Most lease arrangements specify that the first lease payment be made at the start of the lease. The owner specifies to the farm lessee the amount of the payment.

| Obligations on Leased Asset | XXX | |
| Cash | | XXX |

■ At the end of the year, the farm accountant calculates and records depreciation for the leased asset (unless the leased asset is land). The balance sheet reports the book value of the leased asset.
■ At the end of the year, the farm accountant recalculates the amount of the lease liability based on the future payments and calculates the amount of interest paid in the lease payments for the year. The income statement includes this amount of interest. The balance sheet reports the recalculated lease liability.

Table 9-8 outlines the procedures for calculating the lease liability at the end of each year of the lease term and the interest for each year of the lease term.

In Step 1, the farm accountant calculates the present value of the lease payments. The **present value of the lease payments** is the amount of the future payments without the interest. Lease payments include a portion for interest, even though the lease agreement does not specify an interest rate. The owner includes an interest component when calculating the payment amount. If the farm lessee paid for the entire lease up front, the amount paid would be the present value of the payments. The present value of the payments at the start of the lease is most likely the fair market value of the asset (except for salvage value, which the owner excludes from the payments). The **present value factor** (PVF) is a multiplier that discounts the payments to their present value. The factor can be found by using present value tables found in some financial books, by using a financial calculator or computer, or by using a formula, as shown in Table 9-8.

TABLE 9-8 ■ *Calculation of Capital Lease Liability and Annual Interest on Capital Leases.*

Step 1. For the year-end lease liability, compute the present value of the remaining payments. The lease payment is the amount specified in the lease contract, paid by the lessee.

Year-end lease liability = Lease payment × PVF

PVF = present value factor

$= \{1 - [1 \div (1 + i)^n]\} \div i$

n = the total number of payments remaining on the lease

i = annual interest rate ÷ the number of payments made in a year

Step 2. Calculate the change in the lease liability. Subtract the result from Step 1 from the amount of the lease liability reported at the end of the previous year.

Lease liability at the beginning of the year

– Lease liability at the end of the year

= Change in lease liability

Step 3. Calculate the amount of interest for the year.

Interest = Total payments – Change in lease liability

Step 4. An adjustment is required to adjust the liability account to the correct amount and to report interest. In the accrual-adjusted system, the amount from Step 3 is reported as Change in Interest Payable on the income statement and is added to the year-end lease liability amount.

Table 9-8 shows that the formula for the present value factors requires an interest rate (i) and the number of lease payments (n). The farm lessee should estimate an interest rate if the owner does not specify the interest rate used to calculate the lease payment. The estimated interest rate should be a rate that the farm operation would have to pay to borrow a similar amount of money for a similar amount of time. The number of payments made in a year is 12 if monthly payments are made, 4 if quarterly payments are made, or 1 if payments are made only once per year. Each year the total number of remaining payments has to be determined, and this is the value for n. Just as in an installment loan, the amount of each payment and the interest rate remain the same throughout the entire lease term. The only element that changes each year is the number of remaining payments. The result of Step 1 is the current value of the lease liability.

In Step 2, the farm accountant calculates the difference in the value of the lease liability from the end of last year and the current value of the lease liability (calculated in Step1). The change in lease liability is the amount of the lease payments without the interest. Then, in Step 3, the change in lease liability is subtracted from the total amount of lease payments paid during the year. The difference is the amount of the interest included in the lease payments. The adjustment results in the correct amount of interest reported on the income statement and the correct amount of the lease liability. Without the adjustment, the lease liability amount is incorrect because the balance at the end of the year before the adjustment is from the payments, but the payments include some interest. This means that the entire payment did not go toward reducing the liability, so it has to be adjusted back up to the amount still owed on the lease liability.

Suppose that the Farmers enter into a 5-month lease arrangement for some equipment beginning October 1. The owner, an equipment dealer, has agreed to give the Farmers the title to the equipment at the end of the lease term. Because of the ownership transfer, the lease is a capital lease. The fair market value of the leased equipment is $10,000. The owner has decided on monthly payments of $2,040 each. The Farmers estimate an annual interest rate of 12 percent.

At the start of the lease, the Farmers record the following entry to show that they have acquired a lease asset:

Oct. 1, 20X1	1910 Leased Assets	10,000	
	2600 Obligations on Leased Assets		10,000

The following journal entry is recorded for the payment of the first lease payment at the beginning of the lease:

Oct. 1, 20X1	2600 Obligations on Leased Assets	2,040	
	1000 Cash		2,040

The subsequent payments would be recorded in the following way before the end of the year:

Nov. 1, 20X1	2600 Obligations on Leased Assets	2,040	
	1000 Cash		2,040
Dec. 1, 20X1	2600 Obligations on Leased Assets	2,040	
	1000 Cash		2,040

By the end of the year, the schedule of payments looks as follows:

Date	Beginning Balance	Cash Payment	Interest	Reduction of Liability	Ending Balance
Oct. 1, 20X1	$10,000	$2,040	$0	$2,040	$7,960
Nov. 1, 20X1		2,040	0	2,040	5,920
Dec. 1, 20X1		2,040	0	2,040	3,880
Total		$6,120		$6,120	

The ending balance of $3,880 is the lease liability before the adjustment. The lease liability is recalculated for the remaining two payments ($n = 2$) at the 12 percent annual interest rate (i = annual interest rate ÷ the number of payments made in a year = 12% ÷ 12 = 1%):

Year-end lease liability = Lease payment × PVF
$$= \$2,040 \times 1.97040$$
$$= \$4,020 \text{ (rounded)}$$

PVF = present value factor
$$= \{1 - [1 \div (1 + i)^n]\} \div i$$
$$= \{1 - [1 \div (1 + .01)^2]\} \div .01$$
$$= 1.97040$$

Interest expense is the total payments minus the difference between the beginning lease liability and the year-end lease liability:

Change in lease liability = Beginning lease liability − Year-end lease liability
$$= \$10,000 - 4,020$$
$$= \$\ 5,980$$

The total of the payments is $2,040 \times 3 = \$6,120$.

> Interest = Total payments − Change in lease liability
> = $6,120 − 5,980$
> = $ 140

An adjustment is required to adjust the liability account to the correct amount and to report the interest. If the Farmers use journal entries to record adjustments, they would record the following:

| Dec. 31, 20X1 | 8110 Change in Interest Payable | 140 | |
| | 2600 Obligations on Leased Assets | | 140 |

At the end of the lease term, the balance sheet will no longer report a lease liability after the last payment, just like a loan. At the end of the lease term, the balance sheet will no longer report a leased asset. The leased asset will be completely depreciated and returned to the owner, unless the lessee owns it at the end of the lease term. If the lessee owns the asset, the farm accountant reclassifies it from Leased Asset to an appropriate account, such as Machinery and Equipment or Land. The balance sheet reports the value of the asset at its appraised value.

> **PRACTICE WHAT YOU HAVE LEARNED**   *Practice what you have learned by completing Problem 9-7 at the end of the chapter.*

OWNER WITHDRAWALS, OTHER DISTRIBUTIONS, NON-FARM INCOME

The statement of owner equity reports the activities of the owners in the farm business and the total amount of profit earned by the farm business that has not been distributed to the owners. The two main components of the statement of owner equity are Retained Capital and Valuation Equity. Retained Capital consists of the following items:

Learning Objective 7 ■ To review the components of retained capital pertaining to owners and explain the reporting of these items on the statement of cash flows.

- Net Income (amount earned by the farm business in the current year)
- Owner Withdrawals
- Non-Farm Income
- Gifts and Inheritances (Other Distributions)

Table 9-9 shows the example from Appendix A to help you recall the format for the statement of owner equity.

Owner Withdrawals must be accurately recorded or reasonably estimated, regardless of the form of organization of the farm business. The amount should be reported on the statement of cash flows and the statement of owner equity (or net worth) to make comparisons between farm operations more feasible. Owner Withdrawals are distributions from the farm business to cover family living expenses, personal taxes,

TABLE 9-9 ■ *The Farmers' Statement of Owner Equity for 20X1.*

Owners' Equity, Beginning of Period			$104,540.00
Net Income		$40,905.80	
Owner Withdrawals	($150.00)		
Non-farm Income	100.00		
Net Owner Withdrawals		(50.00)	
Gifts and Inheritances		350,000.00	
Addition to Retained Capital			$390,855.80
Valuation Equity			
Change in Excess of Market Value over Cost		18,500.00	
Change in Non-current Deferred Taxes		(6,850.00)	
Total Valuation Equity			11,650.00
Addition to Retained Capital and Valuation Equity			402,505.80
Owners' Equity, End of Period			$507,045.80

purchases of personal assets, and other non-farm items. Owner Withdrawals can be shown in either the operating section or the financing section of the statement of cash flows. Other Distributions should be reported on the statement of cash flows in the financing section. Any Non-Farm Income contributed to the farm business can be shown separately from the owner withdrawals on the statement of owner equity or net worth or can be netted with owner withdrawals. On the statement of cash flows, this item should be shown separately in the financing section. As you learned in Chapter 5, the balances in the Owner Withdrawals, Non-Farm Income, and Gifts and Inheritances accounts are closed to the Retained Capital account, along with Net Income.

VALUATION EQUITY

Learning Objective 8 ■ To explain the purpose of the components of valuation equity.

Valuation equity consists of the effects of market value changes on the farm financial statements. Market value changes are reported separate from Retained Capital because not all farm operations will decide to report market values. Lenders and other parties reading financial statements can more easily compare a farm that reports market values with a farm not reporting market values with this format.

Valuation equity is the sum of two accounts:

■ Change in Excess of Market Value over Cost
■ Change in Non-Current Deferred Taxes.

These two accounts can be shown separately on the balance sheet or combined and shown as Valuation Equity. The computations for these items are discussed throughout this book and are not repeated here.

DISCLOSURE NOTES

Lenders request details about the farm debts when evaluating whether or not to lend money to a farm operation. Common types of information in disclosure notes for this purpose include the following:

Learning Objective 9 ■ To list the information on liabilities disclosed in the notes to the financial statements.

- ■ Disclosure notes for accounts payable and taxes payable should indicate the amounts owed, the parties owed, and the due date.
- ■ Information in the disclosure notes should include all relevant information on all notes payable—name of lender, original principal amount, amount left to pay on the loan, interest rate, due dates of payments, the term of the loan, and the amount of the payments on installment loans.

The farm accountant can create separate Notes Payable accounts for each loan, such as Real Estate Notes Payable shown in Appendix C, to help keep track of payments. Whether one or several Notes Payable accounts are used, the disclosure notes should outline details of the loans according to the arrangements with lenders.

Disclosure notes for deferred taxes should be quite explicit concerning the details of the calculations for current and non-current portions of deferred taxes. Accurate and complete records can be facilitated by disclosing the computations for the current deferred taxes and both components of non-current deferred taxes in the notes to the financial statements each year.

- ■ The first section in the notes should list each of the current assets and their values from the previous and current years so that the computations for Steps 1 and 2 for current deferred taxes are clear.
- ■ The next items to list are the current liabilities and their values from both last year and this year.
- ■ The amount of current deferred tax liability (results of Step 4) should be shown for both years (last year and this year).
- ■ The next section should provide base value and tax basis information for each class of raised breeding livestock (for last year and this year).
- ■ The amount of the first component of non-current deferred taxes related to base values of raised breeding livestock is shown for both years.
- ■ The next section continues with the purchased non-current assets, listing each asset, its book, its tax basis, and the difference between the two for each asset, again providing this information for both last year and this year.
- ■ The next section should provide market values and the book or base values of all non-current assets for both last year and this year.
- ■ The deferred tax liabilities for the second component of non-current deferred taxes for each year are shown.
- ■ The deferred tax liabilities for the second component are added to the deferred tax liabilities for the first component for each year.

Disclosure information for operating leases should include

- ■ Descriptions of all assets involved in capital leases
- ■ The amount of the lease payments
- ■ The interest rate used in each present value
- ■ The number of payments
- ■ Method of depreciation

The computations for the change in lease liabilities, total payments, and interest should be clear from the information in the notes.

The computation for the valuation equity should be explicit in the disclosure notes if not reported on the statement of owner equity.

CHAPTER SUMMARY

Valuation procedures for liabilities such as Accounts Payable, Taxes Payable, and Interest Payable are quite straightforward because the amount of each liability is presented in the form of a bill from some outside party. However, the valuation of other liabilities can involve extensive procedures, such as those for deferred taxes. Deferred tax liabilities should be performed after all adjustments have been made. In many respects, the accounting procedures for capital leases resemble the procedures for installment notes payable. Accounting for capital leases involves the present value computation for the lease liability.

The components of valuation equity should be reported, to accurately present the information related to the non-current deferred taxes and retained capital. The more complete the disclosure information on liability and equity accounts, the more informative and the more useful the financial reports will be. Because of the complex nature of some of these items, the amount of disclosure information is crucial for lending and other decisions that rely on the accuracy of the financial information for the farm business.

PROBLEMS

9-1 ■ For each of the transactions below, identify whether you would report it as an expense (E), a liability (L), or a "change" item (C), the name of the account that you would use to report it, and on which financial statement it would appear.

a. The farm accountant receives the telephone bill and pays the bill immediately.

b. The farm operation owes $310 to the local cooperative at the end of the year.

c. The farm accountant compares the $500 accounts payable owed at the beginning of the year to the $310 owed at the end of the year.

d. The farm accountant pays the bill in (b) after the beginning of the year.

e. The farm accountant pays the feed bill.

f. The farm accountant estimates that the farm operation owes $2,800 in income taxes at the end of the year. The farm accountant compares the $3,100 taxes payable at the beginning of the year to the $2,800 owed at the end of the year.

g. The farm accountant pays the income taxes.

h. The farm accountant estimates that the farm operation owes $5,000 in interest at the end of the year.

i. The farm accountant pays off a loan including interest.

9-2 ■ a. Repeat the procedures for the Farmers' remaining payments on the purchase of the harvester. Verify the amounts for interest expense and reduction of liability by calculating the interest and principal portions for the third, fourth, and fifth payments. Then, prepare the journal entries for these payments.

b. Suppose that the Farmers are making monthly payments, instead of annual payments. Explain how the amount for interest would be calculated differently with monthly payments instead of annual payments. (Do not do the actual calculations.)

9-3 ■ Help the Farmers estimate the current deferred taxes for the year 20X2. Use the information in Chapter 5 for the subsequent year's adjustments to make your calculations.

9-4 ■ Help the Farmers estimate the first component of non-current deferred taxes for the year 20X2. Use the individual animal approach with multiple transfer points and the changes in base values illustrated in Chapter 8 for the raised breeding livestock. Use the same depreciation methods that the Farmers used in Chapter 4.

9-5 ■ Help the Farmers estimate the second component of non-current deferred taxes for the year 20X2. Use the market values below and the book values that you calculated in Problem 9-4.

	Market Values
Breeding Livestock	70,000
Machinery and Equipment	64,000
Office Equipment and Furniture	700
Perennial Crops	42,000
Land	400,000
Buildings/Improvements	100,000
Leased Assets	93,333

9-6 ■ Identify each of the descriptions below as a characteristic of an operating lease (O), or a capital lease (C), or as an installment purchase (I). Some of the descriptions may apply to more than one.

a. Rental arrangements that do not meet the criteria for a capital lease.

b. Special types of leases that, in effect, are installment purchases.

c. The buyer obtains the title of the asset at the time of the purchase and then makes periodic payments.

d. The buyer obtains the title of the asset after making periodic payments.

e. The farm accountant records the rental payments as rent expense.

f. One party continues to have title of an asset while another party uses it, but will transfer the title to the party using it.

g. The party using the asset has all of the benefits and risks of the owner of the asset.

h. The risks of owning an asset remain with the owner while another party uses the asset.

i. The party using the asset can report the asset along with other owned assets on the balance sheet and can record depreciation expense.

j. The party using the asset does not report the asset on the balance sheet and does not record depreciation.

k. The payments for the asset include interest.

l. The payments for the asset do not include interest.

m. The party using the asset reports a liability.

9-7 ■ Calculate the interest on each of the capital leases below.

	Beginning lease liability	Year end lease liability	Total payments
a.	$100,000	$45,020	$64,000
b.	$ 20,000	$14,680	$ 6,120
c.	$ 45,000	$34,900	$ 3,600
d.	$ 10,000	$ 6,020	$ 3,000

Analyzing Financial Position and Financial Performance

I n Chapters 1 through 9, you learned how to prepare accrual-adjusted farm financial statements and how to measure and value revenues, expenses, assets, liabilities, and equity items.

This chapter focuses on evaluating the financial performance and financial position of a farm operation. You will learn about comparative financial statements and how to conduct horizontal and vertical analyses. You will learn the definitions of liquidity, solvency, repayment capacity, profitability, and financial efficiency and how to analyze each of these areas of a farm business. As mentioned in Chapter 1, financial statements should provide feedback on the decisions made by the farm owner or manager. The techniques in this chapter show you how to obtain the feedback from the financial statements.

The accounting procedures outlined in this book provide guidelines that agricultural producers can use as an alternative to GAAP procedures. The financial statements and disclosure notes prepared according to the FFSC Guidelines are intended to evaluate the financial position and financial performance of agricultural operations by producers, lenders, and other interested parties. Financial statements not prepared according to GAAP or the FFSC Guidelines can misstate the financial position or financial performance of a farm operation, and can inhibit comparing one farm to another.

You have learned that one of the primary features of GAAP and the FFSC Guidelines is adhering to reporting revenue when earned and reporting expenses when incurred (the matching concept). Matching is accomplished using the accrual-basis system recommended in GAAP and the accrual-adjusted approach recommended in the FFSC Guidelines. Financial statements prepared under a cash-basis system might not report revenues and expenses in the period in which they occur. Because selling farm products and paying for expenses can occur at different times from year to year, comparisons from one year to the next may be difficult. Financial statements prepared using accrual-basis accounting or the accrual-adjusted approach alleviate that problem by reporting revenues when earned and expenses when occurred, instead of when cash is paid or received. In addition, using the measurement and valuation procedures in the FFSC Guidelines contributes to consistent reporting from one year to the next and from one farm to the next.

Lenders and agricultural producers recognize that other factors besides financial information will play a role in analyzing the health of a farm business. These factors are as varied as the nature of each farm operation, its strengths and weaknesses, and its management. This chapter concerns financial indicators to use in evaluating virtually any farm operation. Although some indicators for evaluating a farm operation might be more important than others, the financial indicators discussed in this chapter are universal in nature. The individual producer or other interested parties decide on the usefulness of each indicator for the farm operation.

COMPARATIVE FINANCIAL STATEMENTS

Learning Objective 1 ▨ To define "comparative financial statements" and conduct a horizontal analysis and trend analysis.

Comparative financial statements provide one way to evaluate the financial health of a farm business. They consist of statements with more than one year of data. The current year's data is presented alongside the previous year's data, as shown in Table 10-1.

When financial statements are prepared for the current year, the financial statements will also include the line items on each statement for the previous year in side-by-side vertical columns. Presenting the information in this manner helps producers and lenders compare the current year's financial position and performance with previous years.

Comparative financial statements can be used for horizontal analysis. **Horizontal analysis** is using percentage changes of financial items to assess improvement. Horizontal analysis involves:

- ▨ Calculating the percentage changes of each item on the financial statements.
  - • subtract the amount from the previous year from the current year's amount
  - • divide the difference by the amount from the previous year
- ▨ Evaluating the results.

Percentage changes of key items can reveal whether or not the financial performance has improved over time. For example, a percentage increase in accounts payable might suggest problems with paying bills. Percentage changes can be useful, but can also be misleading. If net income, for example, is quite low or negative in a given year because of an extraordinary loss, and returns to normal levels the following year, the percentage change will be quite drastic. Readers of the financial statements should understand the nature of individual items, to avoid being misled by such abnormalities.

TABLE 10-1 ■ *Farmers' Partial Comparative Income Statements and Balance Sheets.*

INCOME STATEMENT		
	20X2	**20X1**
Gross Revenues	$79,600.00	$81,050.00
Operating Expenses	(46,800.00)	(31,564.20)
Net Farm Income from Operations	32,800.00	49,485.80
Loss on Sales of Farm Capital Assets	0	(800.00)
Income before Taxes	32,800.00	48,685.80
Total Income Tax Expense (Farm Business Only)	(6,500.00)	(7,780.00)
Accrual Adjusted Net Income	$26,300.00	$40,905.80

BALANCE SHEET					
20X2	**20X1**			**20X2**	**20X1**
Assets:		Liabilities:			
		Total Current Liabilities		$ 45,500.00	$ 59,820.00
Total Current Assets $ 17,800.00	$ 16,403.80	Total Non-Current Liab.		145,000.00	206,168.00
		Total Liabilities		190,500.00	265,988.00
		Equity:			
		Retained Capital		561,700.00	495,615.80
		Valuation Equity		10,950.00	11,650.00
Total Non-Current Assets 745,350.00	756,850.00	Total Equity		572,650.00	507,265.80
Total Assets $763,150.00	$773,253.80	Total Liabilities and Equity		$763,150.00	$773,253.80

The Farmers conducted a horizontal analysis on their income statement after they prepared the financial statements for 20X2. They decided to conduct the analysis on gross revenues, operating expenses, net farm income from operations, income before taxes, total income tax expense, and accrual adjusted net income.

	Year 2 minus Year 1	Divide by Year 1
Percentage changes for: Gross Revenues	= [79,600 − 81,050] = −1.8%	÷ 81,050
Operating Expenses	= [46,800 − 31,564] = +48%	÷ 31,564
Net Farm Operating Income	= [32,800 − 49,486] = −34%	÷ 49,486
Income before Taxes	= [32,800 − 48,686] = −33%	÷ 48,686

Income Tax Expense	= [6,500 − 7,780]		÷	7,780
	= −16%			
Accrual Adjusted Net Income	= [26,300 − 40,906]		÷	40,906
	= −36%			

Income Statement

	20X2	20X1	Percentage Change
Gross Revenues	$79,600.00	$81,050.00	−1.8%
Operating Expenses	(46,800.00)	(31,564.20)	+49%
Net Farm Income from Operations	32,800.00	49,485.80	−34%
Loss on Sales of Farm Capital Assets	0	(800.00)	
Income before Taxes	32,800.00	48,685.80	−33%
Total Income Tax Expense (Farm Business Only)	(6,500.00)	(7,780.00)	−16%
Accrual Adjusted Net Income	$26,300.00	$40,905.80	−36%

Exercise 10-1 *Can you evaluate the results of the Farmers' horizontal analysis? How did the farm business perform in 20X2 compared to 20X1?*

Answer: Clearly, the farm business produced less profit in 20X2 compared to 20X1. The revenues and the net farm operating income, income before taxes, and accrual adjusted net income decreased from the previous year. We can attribute the decrease in these numbers mostly to the increase in operating expenses.

Trend analysis is calculating the percentage change from a base year.

- The amount for each of the line items in the first year's financial statements is the base figure for that item.
- In each subsequent year, each line item on the financial statements is divided by the corresponding line item on the first year's financial statements.

The purpose of the trend analysis is to check the overall trend. Examining the overall trend helps evaluate the general direction of the farm's performance. The percentages indicate whether or not the farm operation is improving financially. This technique can also alleviate distortions that can result from looking at percentage changes only from one year to the next.

Suppose that for the Farmers, their first year of operation is the year 20X1. The amounts for each of the line items in the 20X1 financial statements are the base figures. For accrual-adjusted net income, for example, each subsequent year's accrual-adjusted net income is divided by the accrual adjusted net income for 20X1.

	20X1	20X2	20X3	20X4
Accrual Adjusted Net Income	$40,906	$26,300	$43,685	$47,395

Trend Analysis:	Current Year	÷	20X1
20X2:	= 26,300	÷	40,906
	= 64%		
20X3:	= 43,685	÷	40,906
	= 107%		
20X4:	= 47,395	÷	40,906
	= 116%		

Exercise 10-2 *Can you evaluate the results of the Farmers' trend analysis? How is the farm business performing over time?*

Answer: With the exception of 20X2, accrual adjusted net income is on an upward trend, with 20X3 and 20X4 reporting an increase over 20X1. The increase in 20X4 is greater than the increase in 20X3, further indicating an upward trend.

Horizontal analysis and trend analysis enable comparison of the farm operation over time. For the lender, these techniques allow for comparisons with other farm operations.

PRACTICE WHAT YOU HAVE LEARNED *Practice these techniques by completing Problem 10-1 at the end of the chapter.*

FINANCIAL RATIOS

Typically, a farm operation has obligations to outside parties (creditors). Creditors are interested in measuring the ability of the farm business to pay its debts. The owners of the farm business are interested in measuring the profitability of the farm business and the return on the owners' investment. In addition to horizontal and trend analysis, owners and creditors can use various financial ratios to measure financial performance and financial position. They are interested in different types of ratios that pertain to their particular interests.

A financial ratio is a mathematical relationship of one financial item to another. They are especially useful for comparing information from one farm operation to another farm operation or from one year to another year for a single farm operation. Financial ratios level the playing field for making reasonable comparisons. For an example, see Table 10-2.

Learning Objective 2 ■ To define "liquidity," "solvency," "repayment capacity," "profitability," and "financial efficiency."

TABLE 10-2 ■ *Two Different Sized Farm Operations.*

	Farm A	Farm B
Accrual Adjusted Net Income	$ 50,000	$ 100,000
Total Assets	$400,000	$1,000,000

Farm A reports accrual-adjusted net income of $50,000 and Farm B reports accrual-adjusted net income of $100,000. These numbers by themselves suggest that Farm B is twice as profitable as Farm A. However, if Farm B has assets of $1,000,000 and Farm A has assets of $400,000, clearly Farm A makes more money from its assets. Farm B has 2 1/2 times more assets than Farm A, but only twice the net income.

Financial ratios can be categorized into several groups, each providing specific financial information. Many meaningful ratios can be constructed, depending on the needs of the creditors or owners. The FFSC Guidelines specify five categories of ratios based on performance criteria. Creditors are primarily interested in the first three criteria. Owners are primarily interested in the fourth and fifth criteria but may also be somewhat interested in the other three criteria. These performance criteria are:

- Liquidity
- Solvency
- Repayment Capacity
- Profitability
- Financial Efficiency

Liquidity concerns the ability of a farm business to meet its financial obligations when due. These obligations arise mainly from operating activities and are usually short-term in nature.

Solvency concerns the ability of the farm business to continue to operate as a viable business regardless of external forces, and the ability of the farm business to pay all financial obligations if all assets were sold. Measures of solvency focus primarily on the long-term horizon.

Repayment capacity concerns the ability of the farm business to pay back loans from farm and non-farm income.

Profitability concerns the ability of the farm business to generate a profit.

Financial efficiency concerns the ability of the farm business to use its assets to generate revenues and the effectiveness of its operating, investing, and financing activities.

The ability to use financial ratios to evaluate the financial position and financial performance of a farm business requires an understanding of

- How to calculate each ratio
- How to interpret each ratio
- What benchmark to use to evaluate the results
- The limitations of each ratio

The following discussion outlines these points for each ratio.

PRACTICE WHAT YOU HAVE LEARNED   *At this point, you have learned about what information financial statements and disclosure notes provide. Take the position of a creditor and an owner of a farm business and complete Problem 10-2 at the end of the chapter.*

LIQUIDITY RATIOS

Learning Objective 3 ▪ To analyze the liquidity of a farm business.

Liquidity ratios measure the ability of a farm business to meet its financial obligations when due. They include

- The current ratio
- Working capital

Current Ratio

The **current ratio** measures the ability of the farm business to pay short-term debts. This ratio indicates the extent to which farm current assets can cover farm current liabilities.

How to calculate the current ratio

Current ratio = Total current farm assets ÷ Total current farm liabilities

How to interpret this ratio

The higher this ratio, the less likely the farm producer needs to borrow money to pay bills.

What benchmark to use to evaluate the results

Benchmarks for this ratio include trends of this ratio over time, comparisons of this ratio with ratios from other similar farm operations, or simply the number 1. If the ratio is greater than 1, the current liabilities are less than the current assets available to cover them. However, for some farm businesses, a ratio less than 1 may be typical but the farm business is paying bills on time.

Limitations of this ratio

The limitations of this ratio include the nature of the current assets. For example, suppose a farmer includes a growing crop of wheat as inventory. Including a value for the crop in inventory increases the current assets, increases the numerator in the current ratio, and, therefore, increases the current ratio (see Table 10-3). However, the farmer will not harvest the crop for some time. The cash from selling the crop is not available to pay current liabilities until after the harvest. For this reason, growing crops are not normally included in inventory until after harvest. The ability to convert each current asset into cash must be evaluated when using this ratio.

TABLE 10-3 ▪ *Example of Current Ratios with and without Growing Crop.*

CURRENT ASSETS		CURRENT LIABILITIES	
Cash	$10,000	Accounts Payable	$ 3,000
Inventory	15,000	Interest Payable	16,000
		Taxes Payable	1,000
		Current Deferred Taxes	1,600
Totals	$25,000		$21,600

The value of the growing crop is $60,000.

Current ratio without the growing crop:

Current ratio = Current assets ÷ Current liabilities
 = $25,000 ÷ 21,600
 = 1.16

Current ratio with the growing crop:

Current ratio = Current assets ÷ Current liabilities
 = [$25,000 + 60,000] ÷ 21,600
 = 3.94

Other considerations in the use of this ratio include the following:

- The ratio only indicates the availability of current assets at a given time. The current assets (particularly cash) change frequently. The ratio is not necessarily a predictor of future liquidity.
- This ratio should not be considered in isolation from other types of useful information. The due dates of each liability and cash flow projections would provide additional information.
  - **Cash flow projections** indicate when and how much cash will flow into the farm business and when and how much will flow out to pay bills and service debts.
- The values of current assets and current liabilities may be estimates, particularly the market values of assets, and thus will affect the calculation of the ratio.
- The value of the ratio will vary throughout the production cycle. Therefore, the farm accountant or lender should calculate it at the same time every year or every month and compare it with the same time last year or last month.

Working Capital

Working capital is the amount of funds available after the sale of current farm assets and payment of all current farm liabilities. Working capital is a dollar amount, not a ratio.

How to calculate working capital

Working Capital = Total current farm assets − Total current farm liabilities

How to interpret this number

The higher this number, the less likely that the farm producer will have to borrow money to pay bills.

What benchmark to use to evaluate the results

The most useful benchmark is a comparison with working capital in a previous period at the same point in the production cycle. Comparisons with other farm businesses are futile because of the different size of farm operations. No industry or numerical benchmarks exist for the same reason.

Limitations of working capital

The limitations and considerations of working capital are virtually the same as for the current ratio. Some non-farm business analysts may argue that working capital has limited usefulness, and that the current ratio is a better measure of liquidity.

Using the data in Table 10-1, the Farmers analyze the liquidity of their farm business at the end of 20X1 and 20X2.

	20X2	20X1
Current ratio	$17,800 \div 45,500 = .39$	$16,404 \div 59,820 = .27$
Working capital	$17,800 - 45,500 = (27,700)$	$16,404 - 59,820 = (43,416)$

The current ratio in both years is below 1, but is improving from 20X1 to 20X2. The working capital in both years is negative, but improving also. The Farmers need to examine their cash flow to determine whether they are having difficulty meeting their financial obligations.

PRACTICE WHAT YOU HAVE LEARNED *Practice what you have learned by completing Problem 10-3 at the end of the chapter.*

SOLVENCY RATIOS

Solvency ratios measure the ability of the farm business to continue to operate as a viable business regardless of external forces, and of the ability of the farm business to pay all financial obligations if all assets were sold. Measures of solvency include:

Learning Objective 4 ■ To analyze the solvency of a farm business.

- ■ Debt to assets ratio
- ■ Equity to assets ratio
- ■ Debt to equity ratio

Figure 10-1 shows how these ratios relate to each other.

Debt to Assets Ratio

The **debt to assets ratio** indicates the percentage of assets financed by creditors. It can be calculated using either cost or market values for farm assets. If the market

FIGURE 10-1 ■ *Relationships of Solvency Ratios.*

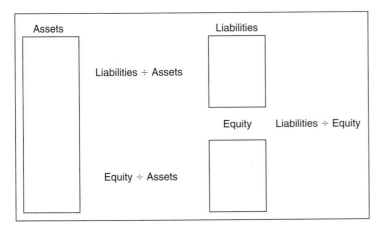

values are used, the second component of deferred taxes should be included as part of the liabilities.

How to calculate the debt to assets ratio

Debt to Assets Ratio = Total farm liabilities ÷ Total farm assets

How to interpret this ratio

The higher the ratio, the greater the credit risk.

What benchmark to use to evaluate the results

Benchmarks include comparisons with previous years for the farm business and comparisons with other similar farm operations. When comparing with other farms, the ratio is most meaningful using market values because cost values may not be available or may vary considerably. When comparing a single farm's position with previous periods, the cost approach is more appropriate and meaningful. No numerical benchmark is meaningful with respect to the debt to assets ratio.

Limitations of this ratio

Market values of farm assets may be highly subjective and may fluctuate. Fluctuations in market values affect the denominator of the ratio, making comparisons difficult from one year to another. Book values are lower for old assets. Denominators may be low if assets are old; however, debts might be lower also if assets are not replaced.

Equity to Assets Ratio

The **equity to assets ratio** indicates the percentage of assets financed by owners. This ratio can be calculated using either cost or market values for farm assets. If the market value approach is used, the deferred taxes related to the assets should be included as part of the liabilities.

How to calculate the equity to assets ratio

Equity to Assets Ratio = Total farm equity ÷ Total farm assets

How to interpret this ratio

The higher the ratio, the lower the credit risk.

What benchmark to use to evaluate the results

The benchmarks of this ratio are virtually the same as for the debt to assets ratio.

Limitations of this ratio

The limitations of this ratio are virtually the same as for the debt to assets ratio.

Debt to Equity Ratio

The **debt to equity ratio** indicates the extent to which creditors have supplied capital relative to the amount of capital supplied by owners and the accumulated results of operations. When combined financial statements are prepared, the debt to net worth ratio is appropriate.

How to calculate these ratios

Debt to Equity Ratio = Total farm liabilities ÷ Total farm equity
Debt to Net Worth = Total liabilities ÷ Total net worth

How to interpret these ratios

The higher this ratio, the greater the credit risk.

What benchmark to use to evaluate the results

Benchmarks include comparisons with previous years for the farm business and comparisons with other similar farm operations. A numerical benchmark of less than 1 (or less than 100 percent) may be useful, depending on the nature of the farm business. A ratio of less than 100 percent indicates that the farm business owes less to creditors than to owners. Thus, owners may be able to recover some assets in the case of liquidation.

Limitations of these ratios

Equity includes valuation equity in market-based financial statements. Market values of farm assets may be highly subjective and may fluctuate. Fluctuations in market values affect the denominator of the ratio, making comparisons difficult from one year to another.

Using the data in Table 10-1, the Farmers analyze the solvency of their farm business at the end of 20X1 and 20X2.

	20X2	20X1
Debt to Assets	$190,500 ÷ 763,150 = 25%	$265,988 ÷ 773,254 = 34%
Equity to Assets	$572,650 ÷ 763,150 = 75%	$507,266 ÷ 773,254 = 66%
Debt to Equity	$190,500 ÷ 572,650 = 33%	$265,988 ÷ 507,266 = 52%

The Farmers' solvency position improved from 20X1 to 20X2, with lower debt to assets and debt to equity ratios. This improvement is due primarily to lower debt. Assets decreased and equity increased, causing the equity to assets ratio to increase. The assets decreased because of depreciation, but it is not clear from the data if market values or book values are used in the ratios. The statement of owner equity would help explain why equity increased. Accrual-adjusted net income decreased, so it is not clear why equity increased while assets decreased.

> **PRACTICE WHAT YOU HAVE LEARNED**   *Practice what you have learned and complete Problem 10-4 at the end of the chapter.*

REPAYMENT CAPACITY RATIOS

Repayment capacity ratios measure the ability of the farm business to pay back loans from farm and non-farm income. These measures include:

■ Term debt and capital lease coverage ratio
■ Capital replacement and term debt repayment margin

Term Debt and Capital Lease Coverage Ratio

The **term debt and capital lease coverage ratio** measures the ability of the farm business to cover all term debt and capital lease payments. This ratio is a measure of the amount of income before interest and after income taxes and owner withdrawals relative to the payments on debt and capital leases.

How to calculate the term debt and capital lease coverage ratio

Term Debt and Capital Lease Coverage Ratio =
[Net farm income from operations
+/− Miscellaneous revenue/expense
+ Non-farm income
+ Depreciation expense
+ Interest on term debt
+ Interest on capital leases
−Total income tax expense
−Owner withdrawals]
÷
[Annual principal and interest payments on term debt
+ Annual payments on capital leases]

How to interpret this ratio

The higher the ratio, the better able the farm business is to make debt and capital lease payments.

What benchmark to use to evaluate the results

Comparisons with previous years or with other similar farm operations provide useful benchmarks. An appropriate numerical benchmark is a ratio of 1 or 100 percent. A ratio equal to 1 indicates that enough cash exists to cover the debt and capital lease payments.

Limitations of this ratio

The appropriate ratio for each farm operation will depend on the type of business, the degree of diversification of the farm operation, and the management abilities of the farmer. This ratio should be considered along with cash flow. The farm business may generate enough earnings to cover all debt and capital lease payments, but sufficient cash must be available to make the payments on time. This ratio makes no provision for the replacement of capital assets. When new debt occurs, the comparison of the ratio with ratios of previous years will be less meaningful. Items such as miscellaneous income/expense, non-farm income, and owner withdrawals may vary considerably from year to year, resulting in variations in this ratio that have little to do with the ability of the farm business to generate enough earnings to cover debt and capital lease payments.

Capital Replacement and Term Debt Repayment Margin

The **capital replacement and term debt repayment margin** measures the ability of the farm producer to generate sufficient funds for repayment of non-current debt and to replace capital assets. It also measures the ability to acquire capital assets or service additional debt. This margin provides an indication of the actual dollar amount of income before interest and after income taxes, owner withdrawals, payments on debt and capital leases due within one year, and annual payments for personal liabilities.

How to calculate the capital replacement and term debt repayment margin

First, calculate capital replacement and term debt repayment capacity

1. Capital replacement and term debt payment capacity =
   Net farm income from operations
   +/− Miscellaneous revenue/expense
   + Non-farm income
   + Depreciation expense
   + Interest on term debt
   + Interest on capital leases
   − Total income tax expense
   − Owner withdrawals

2. Capital replacement and term debt repayment margin =
   Capital replacement and term debt payment capacity
   − Payment on unpaid operating debt from a prior period
   − Principal payments on current portions of term debt
   − Principal payments on current portions of capital leases
   − Total annual payments on personal liabilities (if not included in owner withdrawals)

How to interpret this measure

The higher the margin, the more able the farm business is to make payments on non-current debt and replace capital assets.

What benchmark to use to evaluate the results

The margin should be compared over time to determine the stability of the income stream to support additional debt. The magnitude of the margin will suggest the amount of additional payments that the farm operation may be able to cover.

Limitations of this measure

The appropriate margin for each farm operation will depend on the type of business, the degree of diversification of the farm operation, and the farmer's management abilities. This margin should be considered along with cash flow. The farm business may generate enough earnings to cover all debt and capital lease payments, but sufficient cash must be available to make the payments on time. Items such as miscellaneous income and expense, non-farm income, and owner withdrawals may vary considerably from year to year. That can result in variations in this ratio that have little to do with the ability of the farm business to generate enough earnings to cover debt and capital lease payments, and to replace assets.

Using the income statement and statement of owner equity in Appendix A, the Farmers decide to analyze the repayment capacity for their farm business for 20X1. They purchased land in journal entry (6) in Chapter 3. The annual payments on that debt are $14,095. For illustrative purposes, assume that the Farmers purchased, rather than leased, the harvester, so they have an annual debt payment of $23,982 for the harvester. They do have a capital lease on the equipment, with annual lease payments of $2,040.

Term Debt and Capital Lease Coverage Ratio =

[Net farm income from operations	[$49,486
+/– Miscellaneous revenue/expense	0
+ Non-farm income	100
+ Depreciation expense	22,150
+ Interest on term debt	6,200
+ Interest on capital leases	140
– Total income tax expense	(7,780)
– Owner withdrawals]	(150)]
÷	÷
[Annual principal and interest payments on term debt	[14,095
	23,982
+ Annual payments on capital leases]	2,040]

= $70,146 ÷ 40,117
= 175%

The Farmers paid off a 20X0 loan for $10,000 on March 1, 20X1.

Capital replacement and term debt payment capacity = $70,146

Capital replacement and term debt repayment margin =

Capital replacement and term debt payment capacity	$70,146
– Payment on unpaid operating debt from a prior period	(10,000)
– Principal payments on current portions of term debt	(14,095)
	(23,982)
– Principal payments on current portions of capital leases	(2,040)
– Total annual payments on personal liabilities (if not included in owner withdrawals)	0

= $20,029

The interpretation on these numbers is limited because the Farmers do not have any past data for comparison. However, the Term Debt and Capital Lease Coverage Ratio is greater than 100 percent, indicating that the Farmers are beginning their farm operation with the ability to cover their debts. Whether that level of performance will continue remains to be seen.

PRACTICE WHAT YOU HAVE LEARNED   *Practice what you have learned and complete Problem 10-5 at the end of the chapter.*

PROFITABILITY RATIOS

Profitability ratios measure the ability of the farm business to generate a profit. They include:

Learning Objective 6 To analyze the profitability of a farm business.

- Rate of return on farm assets
- Rate of return on farm equity
- Operating profit margin ratio
- Accrual adjusted net farm income

Rate of Return on Farm Assets

The **rate of return on farm assets** measures the ability of farm operators to earn a reasonable profit (return) on the assets that they control. The ratio is based on income from operations before interest and after owner withdrawals. The ratio is most meaningful when taxes are not included because of changes in tax laws and the sometimes difficult task of separating income taxes on non-farm income from farm income taxes. Interest and taxes are generally not under the control of the farm operator and should be excluded for that reason.

How to calculate rate of return on farm assets

Rate of Return on Farm Assets =
[Net farm income from operations + Farm interest expense
– Owner withdrawals] ÷ Average total farm assets

Average total farm assets =
[Total farm assets at the beginning of the year + Total farm assets at the end of the year] ÷ 2

How to interpret this ratio

The higher the ratio, the greater is the amount of profit per dollar of farm assets.

What benchmark to use to evaluate the results

If used for comparisons between farms, this ratio is most meaningful when using market values for farm assets for the same reasons given for the solvency ratios. If used for comparisons between years for the same farm operation, the ratio is most meaningful when using book values for valuing farm assets because of fluctuations in market values. No numerical benchmarks are useful or meaningful.

Limitations of this ratio

The method used to value farm assets can affect the value of the ratio. Furthermore, the ratio can be manipulated by putting off the purchase of new assets if the value used is book value. Old assets that are mostly depreciated will have a low book value and will result in a higher rate of return than new assets. A reliance on this ratio for a measure of profitability may motivate the farm operator to make decisions to put off purchases of new assets even though they may be necessary. Using the original cost without adjustments for depreciation may alleviate this concern. The value of the ratio can vary with the nature of the farm business, particularly concerning the amount of owned land or other assets, unless leased assets are reported as capital leases. If not reported as capital leases, the leased assets will not show up on the balance sheet and the amount of farm assets will be a low number, resulting in a high ratio.

Rate of Return on Farm Equity

The **rate of return on farm equity** measures the ability of the farm business to generate a profit on owner equity. This ratio is based on income from operations after owner withdrawals. The ratio is most meaningful when taxes are not included because of changes in tax laws and the sometimes difficult task of separating income taxes on non-farm income from farm income taxes.

How to calculate rate of return on farm equity

Rate of Return on Farm Equity =
[Net farm income from operations – Owner withdrawals]
÷ Average total farm equity

Average total farm equity =
[Total farm equity at the beginning of the year + Total farm equity at the end of the year] ÷ 2

How to interpret this ratio

The higher the ratio, the greater the amount of profit per dollar of equity.

What benchmark to use to evaluate the results

The discussion on benchmarks for rate of return on assets also applies to this ratio.

Limitations of this ratio

A low ratio might indicate an unprofitable farm business, but could be the result of a high equity business. A high ratio can indicate a profitable farm business, but could be the result of a low equity business with a lot of debt. This ratio should be considered together with the debt to equity ratio.

Operating Profit Margin Ratio

The **operating profit margin ratio** measures the return per dollar of gross revenue. This ratio is most meaningful when taxes are not included for the reasons stated for the previous ratios.

How to calculate operating profit margin ratio

Operating Profit Margin Ratio =
[Net farm income from operations + Farm interest expense
– Owner withdrawals] ÷ Gross revenues

How to interpret this ratio

The higher the ratio, the higher the likelihood that the farm operation will make a profit for its owners.

What benchmark to use to evaluate the results

No benchmarks exist for the operating profit margin ratio. Comparisons with other similar farm operations may be useful. The amount of gross revenues varies with the size of the farm operation, but, because this measure is a percentage, comparisons can be made between different farm operations. Comparisons with previous periods for a single farm operation provide a useful evaluation of the trend in profitability.

Limitations of this ratio

Accrual-adjusted gross revenue is based on the valuation methods for raised inventory. If valuation procedures are used inconsistently from year to year, the comparisons may be difficult.

Accrual Adjusted Net Farm Income

Accrual adjusted net farm income measures the return from the farm business to the farmer for unpaid labor and management. This measure is most meaningful when taxes are not included for the reasons stated above. The measure also excludes miscellaneous revenues and expenses.

How to calculate accrual adjusted net farm income

Gross Revenue
– Operating Expenses +/– Accrual Adjustments
– Depreciation Expense
– Interest Expense +/– Change in Interest Payable
= Net Farm Income from Operations
+/– Gains/Losses on the Sale of Farm Capital Assets
+/– Gains/Losses Due to Changes in Base Values of Breeding Livestock
= Accrual Adjusted Net Farm Income

How to interpret this number

The higher the number, the higher the profitability of the farm's operating, investing, and financing activities.

What benchmark to use to evaluate the results

Numerical benchmarks do not exist for accrual adjusted net farm income. Because this measure is a number instead of a percentage, the amount will vary depending on the size and nature of the farm operation. Furthermore, corporations will include salary paid to the farm owner as an operating expense, whereas other forms of farm business organization will not. Therefore, comparisons between farm operations are not likely to provide a meaningful interpretation. The most meaningful use of this measure is comparison with previous periods for a single farm business.

Limitations of this ratio

Accrual-adjusted accounting recommended in the FFSC Guidelines, if properly applied, results in revenues reported when earned and matched with the expenses incurred. If these procedures for matching revenues and expenses are not implemented, this measure may be understated or overstated.

Using the data in Table 10-1 and assuming that owner withdrawals was $150 for 20X1 and $1,500 for 20X2, the Farmers analyze the profitability of the farm business for 20X1 and 20X2. From journal entries (1) and (2) in Chapter 3, the farm assets were $116,700 and the equity was $104,540 at the beginning of 20X1. The income statements do not report miscellaneous revenues or expenses, so the income before taxes is equal to the accrual adjusted net farm income. From Appendix A, the interest expense for 20X1 was $6,200. Assume that interest expense was $4,500 in 20X2.

20X2

Average total farm assets
= [Total farm assets at the beginning of the year + Total farm assets at the end of the year] ÷ 2
= [$773,254 + 763,150] ÷ 2
= $768,202

Rate of Return on Farm Assets
= [Net farm income from operations + Farm interest expense − Owner withdrawals] ÷ Average total farm assets
= [$32,800 + 4,500 − 1,500] ÷ 768,202
= 4.7%

Average total farm equity
= [Total farm equity at the beginning of the year + Total farm equity at the end of the year] ÷ 2
= [$507,266 + 572,650] ÷ 2
= $539,958

Rate of Return on Farm Equity
= [Net farm income from operations − Owner withdrawals] ÷ Average total farm equity
= [$32,800 − 1,500] ÷ 539,958
= 5.8%

Operating Profit Margin Ratio
= [Net farm income from operations + Farm interest expense − Owner withdrawals] ÷ Gross revenues
= [$32,800 + 4,500 − 1,500] ÷ 79,600
= 45%

Accrual Adjusted Net Farm Income = $32,800

20X1

Average total farm assets
= [Total farm assets at the beginning of the year + Total farm assets at the end of the year] ÷ 2
= [$116,700 + 773,254] ÷ 2
= $444,977

Rate of Return on Farm Assets
= [Net farm income from operations + Farm interest expense − Owner withdrawals] ÷ Average total farm assets
= [$49,486 + 6,200 − 150] ÷ 444,977
= 12.5%

Average total farm equity
= [Total farm equity at the beginning of the year + Total farm equity at the end of the year] ÷ 2
= [$104,540 + 507,266] ÷ 2
= $305,903

Rate of Return on Farm Equity
= [Net farm income from operations − Owner withdrawals] ÷ Average total farm equity
= [$49,486 − 150] ÷ 305,903
= 16.1%

Operating Profit Margin Ratio
= [Net farm income from operations + Farm interest expense − Owner withdrawals] ÷ Gross revenues
= [$49,486 + 6,200 − 150] ÷ 81,050
= 69%

Accrual Adjusted Net Farm Income = $48,686

	20X2	20X1
Rate of Return on Farm Assets	4.5%	12.5%
Rate of Return on Farm Equity	5.8%	16.1%
Operating Profit Margin Ratio	45%	69%
Accrual Adjusted Net Farm Income	$32,800	$48,686

Profitability measures decreased substantially from 20X1 to 20X2, mainly due to the increase in assets. The comparison between years is not useful because 20X1 was the first year of the Farmers' operation, during which they acquired a large amount of assets. Comparisons in the future will be more informative.

> **PRACTICE WHAT YOU HAVE LEARNED**   *Practice what you have learned and complete Problem 10-6 at the end of the chapter.*

FINANCIAL EFFICIENCY RATIOS

Financial efficiency ratios measure the effectiveness of operating, investing, and financing activities of the farm business and the ability of the farm business to use its assets to generate revenues. These ratios include:

Learning Objective 7 ■ To analyze the financial efficiency of a farm business.

- ■ Asset turnover ratio
- ■ Operating expense ratio
- ■ Depreciation expense ratio
- ■ Interest expense ratio
- ■ Net farm income from operations ratio

Asset Turnover Ratio

The **asset turnover ratio** measures the level of efficiency in using farm assets to generate farm revenue.

How to calculate the asset turnover ratio

Asset Turnover Ratio = Gross revenues ÷ Average total farm assets

Average total farm assets =

[Total farm assets at the beginning of the year + Total farm assets at the end of the year] ÷ 2

How to interpret this ratio

The higher the ratio, the more efficient the use of farm assets to generate revenues.

What benchmark to use to evaluate the results

Numerical benchmarks do not exist for this ratio. Comparisons with other similar farm operations may be useful. The amount of gross revenues varies with the size of the farm operation, but, because this measure is a percentage, meaningful comparisons can be made. Comparisons with ratios of previous periods for a single farm operation can be useful for measuring progress of the farm business.

Limitations of this ratio

The method used to value farm assets can affect the value of the ratio. Furthermore, the ratio can be manipulated by putting off the purchase of new assets. Old assets that are mostly depreciated will have a low value and will result in a higher turnover. A reliance on this ratio for a measure of efficiency may motivate the farm operator to make decisions to put off purchases of new assets even though they may be necessary. Using original cost for the values (without adjustments for depreciation) may alleviate this concern. The value of the ratio can vary with the nature of the farm business, particularly concerning the amount of owned land or other assets, unless leased assets are reported as capital leases.

DuPont Analysis System

The DuPont Analysis System refers to the relationship of three ratios to measure overall financial performance. The analysis is also useful for evaluating strategies to improve farm financial performance.[1] The first ratio is the operating profit margin ratio. As you learned in this chapter, this ratio is a measure of profitability. It reflects the results of operating decisions relating to costs, yields, sales of farm products, and other operating decisions.[2] A farm manager can evaluate the effect of changes in any of the components of the operating profit margin ratio on overall financial performance. The second ratio in the DuPont analysis is the asset

---

1. Boehlje, Michael, 1994. Evaluating Farm Financial Performance. *Journal of the American Society of Farm Managers and Rural Appraisers* 58: 109–115.
2. Barnard, Freddie L., and Michael Boehlje. 2004. Using Farm Financial Standards Council Recommendations in the Profitability Linkage Model: The ROA Dilemma. *Journal of the American Society of Farm Managers and Rural Appraisers* 68: 7–11.

turnover ratio, which, as you learned, measures the revenue generated from the farm assets. Management decisions affecting this ratio include yields, selling prices, animal productivity (for example, fertility rates, pounds weaned, and so on), and investments in machinery and other non-current assets.[3] When the operating profit margin ratio is multiplied times the asset turnover ratio, the result is the rate of return on assets.

Rate of Return on Assets = Operating Profit Margin Ratio
× Asset Turnover Ratio

The rate of return on assets relates to the rate of return on equity when the rate of return on assets is multiplied by the financial leverage ratio.[4] The financial leverage ratio is the third ratio in the DuPont analysis. It is computed as follows:

Financial Leverage = Average Total Farm Assets ÷ Average Total Farm Equity

Management decisions affecting debt and equity will affect the financial leverage ratio and ultimately the overall financial performance of the farm business.[5] The rate of return on equity is as follows:

Rate of Return on Equity = Rate of Return on Assets × Financial Leverage
= Operating Profit Margin Ratio × Asset
Turnover Ratio × Financial Leverage

However, this computation holds only if interest expense is not added in the operating profit margin ratio.[6] See Figure 10-2 for an illustration.

FIGURE 10-2 ■ *Relationships between Ratios.*

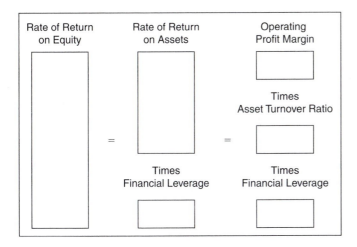

---

3. Ibid.
4. Ibid.
5. Ibid.
6. Ibid.

Using the data in Table 10-1, the Farmers analyze the asset turnover ratios of their farm business for 20X1 and 20X2. They also check the relationship between ratios illustrated in Figure 10-2.

20X2

Operating Profit Margin Ratio (without adding interest)
= [Net farm income from operations − Owner withdrawals] ÷ Gross revenues
= [$32,800 − 1,500] ÷ 79,600
= 39.3%

Average total farm assets
= [Total farm assets at the beginning of the year + Total farm assets at the end of the year] ÷ 2
= [$773,254 + 763,150] ÷ 2
= $768,202

Asset Turnover Ratio
= Gross revenues ÷ Average total farm assets
= $79,600 ÷ 768,202
= 10.4%

Financial Leverage
= Average Total Farm Assets ÷ Average Total Farm Equity
= $768,202 ÷ 539,958
= 142%

Rate of Return on Equity
= Rate of Return on Assets × Financial Leverage
= Operating Profit Margin Ratio × Asset Turnover Ratio × Financial Leverage
= .393 × .101 × 1.46
= 5.8%

20X1

Operating Profit Margin Ratio (without adding interest)
= [Net farm income from operations − Owner withdrawals] ÷ Gross revenues
= [$49,486 − 150] ÷ 81,050
= 60.9%

Average total farm assets
= [Total farm assets at the beginning of the year + Total farm assets at the end of the year] ÷ 2
= [$116,700 + 773,254] ÷ 2
= $444,977

Asset Turnover Ratio
= Gross revenues ÷ Average total farm assets
= $81,050 ÷ 444,977
= 18.2%

Financial Leverage
= Average Total Farm Assets ÷ Average Total Farm Equity
= $444,977 ÷ 305,903
= 145%

Rate of Return on Equity
= Rate of Return on Assets × Financial Leverage
= Operating Profit Margin Ratio × Asset Turnover Ratio × Financial Leverage
= .611 × .182 × 1.45
= 16.1%

The asset turnover ratio decreased significantly from 20X1 to 20X2. The comparison between years is not useful because 20X1 was the first year of the Farmers' operation, during which they acquired a large amount of assets, affecting the average total assets. Comparisons in the future will be more informative.

Operational Ratios

The **operating expense ratio**, the **depreciation expense ratio**, and the **interest expense ratio** measure the percentage of gross revenue used for operating, depreciation, and interest expenses. The **net farm income from operations ratio** is a percentage of profit from farm operations relative to gross revenue. The ratios are most meaningful when taxes are not included in the computations for the reasons indicated for the rate of return on assets.

How to calculate the operational ratios

Operating Expense Ratio = [Total operating expenses – Depreciation
     expense] ÷ Gross revenues
Depreciation Expense Ratio = Depreciation expense ÷ Gross revenues
Interest Expense Ratio = Total farm interest expense ÷ Gross revenues
Net Farm Income from Operations Ratio = Net farm income from
     operations ÷ Gross revenues

Total operating expenses in the operating expense equation does not include interest expense. If interest expense is part of total operating expenses, the total interest expense (including adjustments) must be subtracted from operating expenses in the operating expense ratio.

How to interpret the operational ratios

The lower the expense ratios and the higher the net farm income from operations ratio, the higher the financial efficiency.

What benchmark to use to evaluate the results

No numerical benchmarks exist for any of the ratios. Comparisons with other similar farm operations may be useful if similar accounting procedures are used. However, comparisons with previous periods for a single farm operation may be the most meaningful application of these ratios.

Limitations of each ratio

Accrual-adjusted accounting, if properly applied, results in revenues matched with the expenses incurred. If these procedures for matching revenues and expenses are not consistently applied, comparisons from one year to the next may not be meaningful. Because different depreciation methods may be used, comparisons between farms may be further compromised.

Using the income statement in Table 10-1, the Farmers analyze the financial efficiency of the farm business for 20X1 and 20X2 using operational ratios. The Farmers reported $22,150 of depreciation expense each year. The interest expense for 20X1 was $6,200. Assume that interest expense was $4,500 in 20X2. Depreciation expense and interest expense are included in total operating expenses.

20X2

Operating Expense Ratio
= [Total operating expenses − Depreciation expense − Interest expense] ÷ Gross revenues
= [$46,800 − 22,150 − 4,500] ÷ 79,600
= 25%

Depreciation Expense Ratio
= Depreciation expense ÷ Gross revenues
= $22,150 ÷ 79,600
= 28%

Interest Expense Ratio
= Total farm interest expense ÷ Gross revenues
= $4,500 ÷ 79,600
= 5.7%

Net Farm Income from Operations Ratio
= Net farm income from operations ÷ Gross revenues
= $32,800 ÷ 79,600
= 41%

20X1

Operating Expense Ratio
= [Total operating expenses − Depreciation expense − Interest expense] ÷ Gross revenues
= [$31,394 − 22,150 − 6,200] ÷ 81,050
= 3.8%

Depreciation Expense Ratio
= Depreciation expense ÷ Gross revenues
= $22,150 ÷ 81,050
= 27%

Interest Expense Ratio
= Total farm interest expense ÷ Gross revenues
= $6,200 ÷ 81,050
= 7.6%

Net Farm Income from Operations Ratio
= Net farm income from operations ÷ Gross revenues
= $49,486 ÷ 81,050
= 61%

	20X2	20X1
Operating Expense Ratio	25%	3.8%
Depreciation Expense Ratio	28%	27%
Interest Expense Ratio	5.6%	7.6%
Net Farm Income from Operations Ratio	41%	61%

The operating expense ratio increased substantially, due to lower gross revenues and higher operating expenses in 20X2 compared to 20X1. Efficiency in operating activities needs to improve in the future. The interest expense ratio decreased somewhat, indicating improvement in the financing activities. A complete set of comparative financial statements with percentage changes for each item on the income statements would help to explain the large difference between the two years.

PRACTICE WHAT YOU HAVE LEARNED *Practice what you have learned and complete Problem 10-7 at the end of the chapter.*

CHAPTER SUMMARY

This chapter discusses the purpose and usefulness of various financial measures of financial position and financial performance for farm operations. These measures are useful only with reliable information consistently reported and with appropriate accounting procedures. Financial measures have several limitations and should be considered along with other non-financial measures.

The measures discussed in this chapter are based only on information presented in financial statements using financial accounting procedures. Many other measures of performance that may be more meaningful for farm producers can also be generated. These measures of performance are the basis for management accounting, which is beyond the scope of this book. Nevertheless, the indicators discussed in this chapter can be a useful starting point in evaluation of a farm business, if the limitations of the measures and the measuring processes are considered.

BIBLIOGRAPHY

Barnard, Freddie L., and Michael Boehlje. 2004. Using Farm Financial Standards Council Recommendations in the Profitability Linkage Model: The ROA Dilemma. *Journal of the American Society of Farm Managers and Rural Appraisers* 68: 7–11.

Boehlje, Michael. 1994. Evaluating Farm Financial Performance. *Journal of the American Society of Farm Managers and Rural Appraisers* 58: 109–115.

Farm Financial Standards Council. *Financial Guidelines for Agricultural Producers.* Naperville, IL, 1997. Also available online at http://www.ffsc.org.

PROBLEMS

10-1 ▪ Conduct a horizontal analysis on the Farmers' balance sheet in Table 10-1. Using the data below, conduct a trend analysis on their balance sheet. How would you evaluate the results?

	20X1	20X2	20X3	20X4
Total Current Assets	$ 16,403	$ 17,800	$ 20,100	$ 18,600
Total Assets	773,253	763,150	750,650	768,200
Total Current Liabilities	59,820	45,500	38,600	28,500
Total Liabilities	265,988	190,500	180,900	240,000
Total Equity	507,265	572,650	569,750	528,200

10-2 ▪ a. Take the position of a creditor and list the information that you think you might be interested in to evaluate the liquidity, solvency, and repayment capacity of a farm business.

  b. Take the position of an owner of a farm business and list the information that you think you might be interested in to evaluate the profitability and financial efficiency of your farm business.

10-3 ▪ Using the data in Problem 10-1, analyze the liquidity of the Farmers' business in 20X3 and 20X4. Include an interpretation of your ratio calculations.

10-4 ▪ Using the data in Problem 10-1, analyze the solvency of the Farmers' business in 20X3 and 20X4. Include an interpretation of your ratio calculations.

10-5 ▪ Comment on the repayment capacity of the Farmers' business below.

	20X1	20X2	20X3
Term Debt and Capital Lease Coverage Ratio	175%	180%	165%
Capital replacement and term debt			
repayment margin	$20,249	$30,500	$18,400

10-6 ▪ Using the data in Problem 10-1 and the data below, analyze the profitability of the Farmers' business for 20X3 and 20X4. Include an interpretation of your ratio calculations.

	20X3	20X4
Net farm income from operations	$38,600	$35,800
Farm interest expense	4,500	6,000
Owner withdrawals	1,000	1,200
Gross revenues	82,100	85,300
Accrual adjusted net farm income	38,600	35,000

10-7 ▪ Using the data from Problem 10-1 and the data below, analyze the financial efficiency of the Farmers' business for 20X3 and 20X4. Include an interpretation of your ratio calculations.

	20X3	20X4
Net farm income from operations	$38,600	$35,800
Farm interest expense	4,500	6,000
Gross revenues	82,100	85,300
Total operating expenses	49,300	51,600
Depreciation expense	22,150	27,200

Steve and Chris Farmer
FARM BUSINESS STATEMENT OF INCOME
FOR THE PERIOD ENDING DECEMBER 31, 20X1

Cash Crop Sales		$12,600.00
Change in Crop Inventories		8,300.00
Cash Sales of Market Livestock		50,900.00
Loss from Sale of Culled Breeding Livestock		(250.00)
Change in Value Due to Change in Quantity of Raised Breeding Livestock		6,000.00
Change in Accounts Receivable		3,500.00
Gross Revenues		81,050.00
Operating Expenses:		
Feeder Livestock	(1,500.00)	
Changes in Purchased Feeder Livestock Inventory	750.00	
Purchased Feed	(1,000.00)	
Changes in Purchased Feed Inventory	230.00	
Wage Expense	(1,200.00)	
Payroll Tax Expense	(129.20)	
Truck and Machinery Hire	(150.00)	
Herbicides, Pesticides	(500.00)	
Livestock Supplies, Tools, and Equipment	(75.00)	
Insurance	(1,200.00)	
Real Estate and Personal Property Taxes	(1,300.00)	
Depreciation Expense	(22,150.00)	
Change in Accounts Payable	(340.00)	
Change in Prepaid Insurance	700.00	
Change in Investment in Growing Crop	2,500.00	
Interest Expense	(1,200.00)	
Change in Interest Payable	(5,000.00)	
Total Expenses		(31,564.20)
Net Farm Income from Operations		49,485.80
Loss on Sales of Farm Capital Assets		(800.00)
Income Before Taxes		48,685.80
Income Tax Expense	(2,160.00)	
Change in Taxes Payable	(5,620.00)	
Total Income Tax Expense (Farm Business Only)		(7,780.00)
Accrual Adjusted Net Income		$40,905.80

Learning Objective ■ To identify the elements of the financial statements in each of these expanded formats.

Steve and Chris Farmer
STATEMENT OF OWNER EQUITY
FOR THE PERIOD ENDING DECEMBER 31, 20X1

Owners' Equity, Beginning of Period			$104,540.00
Net Income		$ 40,905.80	
Owner Withdrawals	($150.00)		
Non-Farm Income	100.00		
Net Owner Withdrawals		(50.00)	
Gifts and Inheritances		350,000.00	
Addition to Retained Capital			$390,855.80
Valuation Equity			
Change in Excess of Market Value over Cost		18,500.00	
Change in Non-Current Portion of Deferred Taxes		(6,850.00)	
Total Valuation Equity			11,650.00
Addition to Retained Capital and Valuation Equity			402,505.80
Owners' Equity, End of Period			$507,045.80

Steve and Chris Farmer
BALANCE SHEET
AS OF DECEMBER 31, 20X1

Assets:		Liabilities:	
Cash	$ 203.80	Accounts Payable	$ 340.00
Accounts Receivable	3,500.00	Taxes Payable	3,030.00
		Interest Payable	5,000.00
Feeder Livestock Purchased for Resale	750.00	Notes Payable due within one year	50,000.00
Feed Inventory Raised for Use	2,300.00	Current Deferred Taxes	1,450.00
Feed Inventory Purchased for Use	230.00	Total Current Liabilities	59,820.00
Crop Inventory Raised for Sale	6,000.00		
Prepaid Expenses	700.00	Real Estate Notes Payable	120,000.00
Cash Investment in Growing Crop	2,500.00	Noncurrent Deferred Taxes	10,150.00
Total Current Assets	16,183.80	Obligations on Leased Assets	76,018.00
		Total Non-Current Liab.	206,168.00
Breeding Livestock	64,500.00		
Machinery and Equipment	68,500.00	Total Farm Business Liabilities	265,988.00
Office Equipment and Furniture	1,000.00		
Perennial Crops	45,000.00		
Land, Buildings, Improvements	500,000.00	Equity:	
Leased Assets	100,000.00	Retained Capital	495,395.80
Less: Accumulated Depreciation	(22,150.00)	Valuation Equity	11,650.00
Total Non-Current Assets	756,850.00	Total Equity	507,045.80
Total Farm Business Assets	$773,033.80	Total Liabilities and Owners' Equity	$773,033.80

Steve and Chris Farmer
STATEMENT OF FARM BUSINESS CASH FLOWS
FOR THE PERIOD ENDING DECEMBER 31, 20X1

Cash received from sale of livestock (other than culled breeding livestock)	$ 50,900.00
Cash received from sale of crops	12,600.00
Cash paid for feeder livestock	(1,500.00)
Cash paid for all other operating expenses	(3,943.00)
Cash paid for interest	(1,200.00)
Cash paid for taxes	(3,771.20)
Net cash provided by operating activities	53,085.80
Cash received from the sale of breeding livestock	2,150.00
Cash paid for purchase of machinery and equipment	(69,982.00)
Cash paid for purchase of land and buildings and improvements	(120,000.00)
Cash paid for purchase of investments	(45,000.00)
Net cash used by investing activities	(232,832.00)
Proceeds from real estate and other term loans	170,000.00
Cash received from contributions by owners	20,100.00
Principal payments for loans	(10,000.00)
Owner withdrawals	(150.00)
Net cash provided by financing activities	179,950.00
Net increase in cash from operating, investing, and financing activities	203.80
Cash balance at beginning of year	0
Cash balance at end of year	$ 203.80

Answer to Exercise 1-2:

Revenues: Cash Crop Sales, Change in Crop Inventories, Cash Sales of Market Livestock, Change in Market Livestock and Poultry Inventories, Change in Accounts Receivable.

Expenses: Feeder Livestock, Purchased Feed, Wage Expense, Payroll Tax Expense, Truck and Machinery Hire, Herbicides, Pesticides, Livestock Supplies, Tools and Equipment, Insurance, Real Estate and Personal Property Taxes, Depreciation Expense, Change in Accounts Payable, Change in Prepaid Insurance, Change in Investment in Growing Crop, Interest Expense, Change in Interest Payable, Income Tax Expense, Change in Taxes Payable.

Gains: Change in Value Due to Change in Quantity of Raised Breeding Livestock.

Losses: Loss from Sale of Culled Breeding Livestock, Loss on Sales of Farm Capital Assets.

Answer to Exercise 1-3:

Net income, Investments = Non-farm income + Gifts and Inheritances = $350,100, Withdrawals = $150, and Valuation Equity = Change in Excess of Market Value over Cost and Change in Non-Current Portion of Deferred Taxes.

Answer to Exercise 1-4:

Assets: Cash, Accounts Receivable, Feeder Livestock Inventory Raised for Sale, Feeder Livestock Purchased for Resale, Feed Inventory Raised for Use, Feed Inventory Purchased for Use, Crop Inventory Raised for Sale, Prepaid Expenses, Cash Investment in Growing Crop, Breeding Livestock, Machinery and Equipment, Office Equipment and Furniture, Perennial Crops, Land, Buildings, Improvements, Leased Assets.

Liabilities: Accounts Payable, Taxes Payable, Interest Payable, Notes Payable due within one year, Current Deferred Taxes, Real Estate Notes Payable, Non-current Deferred Taxes, Obligations on Leased Assets.

Equity: Retained Capital, Valuation Equity.

Total assets = $773,033.80.   Total Liabilities and Owners' Equity = $773,033.80.

Answer to Exercise 1-5:

Net cash flows from operating activities = $53,085.80. Net cash used by investing activities = ($232,832.00). Net cash provided by financing activities = $179,950.00.

APPENDIX B

Combined Financial Statements

One of the complex issues concerning farm financial statements is the recording and reporting of personal assets and personal liabilities. Generally, financial statements report the results of activities of a business, but sometimes farm operations prepare combined financial statements that include personal assets and personal liabilities of the farm family on the Balance Sheet. The Balance Sheet displays these items in separate accounts, such as Personal Assets, Personal Liabilities, and Personal Equity.

In many family-owned farm businesses, personal assets and personal liabilities become co-mingled with farm assets and farm liabilities. For example, a purchase for the farm may be charged on a personal credit card and paid all at once along with the personal charges. The farm owner may use the family car to conduct business for the farm operation or might combine non-farm income with farm income. This co-mingling of activities often makes it difficult to separate personal expenses from farm expenses.

When the amount of co-mingled personal expenses and personal income is substantial, an analysis cannot accurately evaluate the financial position and financial performance of the farm operation. The agricultural lender and other outside parties have an interest in understanding the performance of the farm operation separate from the personal financial activities. Yet, it has been a widespread practice to combine the personal financial activities with the farm activities because many farm families have had to rely on non-farm sources of income to assist with farm expenses and debt repayments.

The Farm Financial Standards Council (FFSC) strongly discourages the co-mingling of personal assets and personal liabilities with the farm assets and farm liabilities. It is preferred to have separate sets of financial statements: one set reporting solely the activities of the farm operation and the other to report the personal income, expenses, assets, liabilities, and equity. If this is not possible, then personal assets and personal liabilities should be listed separately on the Balance Sheet from farm assets and farm liabilities, and a detailed list of these assets and liabilities should be included in the Disclosure by Notes. Non-farm income and expenses should not be included in the farm Income Statement and should be shown on the Statement of Owner Equity as a separate item from the farm income. If combined statements are prepared, farm business-only statements should also be prepared.

Table A-1 shows a Balance Sheet for both personal and farm business assets and liabilities.

If both personal and farm financial statements are prepared and if income tax expense can be calculated separately for personal income and farm income, then the farm income tax expense should be shown on the farm Income Statement and the personal income tax expense should be shown on the Statement of Owner Equity. When separation is not feasible, all income tax expense should be reported on the Statement of Owner Equity.[1]

Learning Objective ■ To explain the difference between financial statements of a farm business versus combined financial statements.

---

1. Farm Financial Standards Council. *Financial Guidelines for Agricultural Producers.* (Naperville, IL: 1997).

TABLE A-1 ■ *Balance Sheet Format with Personal and Farm Business Assets and Liabilities.*

Assets	Liabilities
Cash	Accounts Payable
+ Accounts Receivable	+ Taxes Payable
+ Inventory Raised for Sale	+ Interest Payable
+ Inventory Raised for Use	+ Notes Payable due within one year
+ Inventory Purchased for Resale	+ Real Estate debt due within one year
+ Inventory Purchased for Use	+ Current Deferred Taxes
+ Prepaid Expenses	= Total Current Liabilities
+ Cash Investment in Growing Crops	
= Total Current Assets	
Breeding Livestock	Notes Payable – Non-Current
+ Machinery and Equipment	+ Real Estate Note Payable – Non-Current
+ Buildings and Improvements	+ Non-Current Deferred Taxes
+ Perennial Crops, Orchards, and Natural Resources	= Total Non-Current Liabilities
+ Land	Total Business Liabilities
+ Investments in Cooperatives and other Investments	+ Personal Liabilities
= Total Non-Current Assets	= Total Liabilities
	Equity:
Total Business Current and Non-Current Assets	Valuation Equity
+ Personal Assets	+ Retained Capital
= Total Assets	= Total Business Equity
	+ Personal Equity
	= Net Worth
	Total Liabilities and Net Worth

Source: Adapted from Farm Financial Standards Council. *Financial Guidelines for Agricultural Producers* (Naperville, IL: 1997).

A Numbering System for a Chart of Accounts

The FFSC is recommending that all asset accounts begin with the number 1, all liability accounts begin with the number 2, all equity accounts begin with the number 3, revenue accounts begin with 4, production expenses with 5 and 6, selling and general and administrative expenses with 7, other income, other expenses, gains and losses with 8, and income taxes and extraordinary items with 9.[1] Each account is assigned a three- or four-digit number, beginning with the appropriate initial number for their category. The account for cash is usually listed first in a chart of accounts and is assigned the number "100" or "1000." Table A-2 displays this numbering system.

Based on the formats for the four financial statements presented in Tables 1-3, 1-4, 1-5, and 1-6 in Chapter 1, a Chart of Accounts could look something like that depicted in Table A-3. The balance sheet accounts are the 1s, 2s, and 3s. The statement of owner's equity reports details of the 3's and the income statement accounts are the 4s, 5s, 6s, 7s, 8s, and 9s.

TABLE A-2 ■ *FFSC Numbering System for Farm Business Accounts.*

Assets	Liabilities	Equity	Revenues	Expenses	Gains/Losses	Income Taxes
1000	2000	3000	4000	5000	8000	9000
1100	2100	3100	4100	5100	8100	9100
1200	2200	3200	4200	etc.		
etc.	etc.	etc.	etc.	6000		
				6100		
				etc.		
				7000		
				7100		
				etc.		

Source: Adapted from Farm Financial Standards Council. *Financial Guidelines for Agricultural Producers.* (Naperville, IL: 1997).

---

1. Farm Financial Standards Council. *Financial Guidelines for Agricultural Producers.* (Naperville, IL: 1997).

TABLE A-3 ■ *Detailed Farm Chart of Accounts.*

1000 Cash	3100 Retained Capital
1100 Accounts Receivable	3110 Owner Withdrawals
1211 Feeder Livestock Inventory Raised for Sale	3120 Non-Farm Income
1212 Feeder Livestock Inventory Purchased for Resale	3130 Other Capital Contributions/Gifts/Inheritances
1221 Feed Inventory Raised for Sale	3210 Change in Value of Personal Assets
1222 Feed Inventory Purchased for Resale	3220 Change in Personal Liabilities
1223 Feed Inventory Raised for Use	3230 Personal Equity
1224 Feed Inventory Purchased for Use	4000 Cash Crop Sales
1231 Crop Inventory Raised for Sale	4010 Changes in Crop Inventories
1232 Crop Inventory Purchased for Resale	4100 Cash Sales of Market Livestock and Poultry
1300 Prepaid Expenses	4110 Changes in Market Livestock and Poultry Inventories
1400 Cash Investment in Growing Crops	4200 Livestock Products Sales
1500 Breeding Livestock	4300 Proceeds from Government Programs
1510 Breeding Livestock Inventory	4400 Crop Insurance Proceeds
1600 Machinery and Equipment	4500 Gains (Losses) from Sale of Culled Breeding Livestock
1650 Office Furniture and Equipment	4600 Change in Value Due to Change in Quantity of Raised Breeding Livestock
1700 Perennial Crops and Natural Resources	
1800 Land, Buildings, and Improvements	4700 Change in Accounts Receivable
1900 Investments in Cooperatives and other Investments	5000 Feeder Livestock
	5010 Change in Purchased Feeder Livestock Inventories
1910 Leased Assets	5020 Purchased Feed
1950 Personal Assets	5030 Change in Purchased Feed Inventories
1980 Accumulated Depreciation	6100 Wages Expense
2000 Accounts Payable	6110 Payroll Tax Expense
2100 Taxes Payable	6120 Board for Hired Labor
2200 Interest Payable	6130 Insurance for Hired Labor
2300 Notes Payable—Non-Current	6200 Repairs and Maintenance for Farm Vehicles, Machinery, Equipment
2310 Notes Payable Due within One Year	
2400 Real Estate Notes Payable—Non-Current	6210 Small Tools and Supplies
2410 Real Estate Notes Payable Due within One Year	6220 Repairs and Maintenance for Buildings and Improvements
2500 Non-Current Deferred Taxes	6300 Rent
2510 Current Deferred Taxes	6310 Truck and Machinery Hire
2600 Obligations on Leased Assets	6400 Fuel, Oil, Gas, Grease
2610 Obligations on Leased Assets Due within One Year	6500 Seed
	6510 Fertilizers
2700 Personal Liabilities	6520 Herbicides, Pesticides
3000 Valuation Equity	6530 Twine, Sacks
3010 Change in Excess of Market Value over Cost	6540 Poisons, Seed Tests
3020 Change in Non-Current portion of Deferred Taxes	6600 Veterinarian, Vaccinations, Medications

6610 Breeding fees, Registrations

6620 Disinfectants, Sprays

6630 Livestock Supplies, Tools, and Equipment

6640 Shearing

6641 Wool Twine and Sacks

6650 Livestock Inspections

6700 Insurance

6710 Real Estate and Personal Property Taxes

6720 Electricity

6730 Water

6740 Telephone

6750 Office Supplies

6760 Dues, Journals and Papers

6770 Bank Charges

6780 Depreciation Expense

6810 Change in Accounts Payable

6820 Change in Prepaid Insurance

6830 Change in Investment in Growing Crops

7100 Sales Costs

7200 General and Administrative Expenses

8000 Interest Income

8100 Interest Expense

8110 Change in Interest Payable

8200 Gains (Losses) on Sales of Farm Capital Assets

8300 Gains (Losses) Due to Changes in General Base Values of Breeding Livestock

8400 Miscellaneous Revenue

9100 Income Tax Expense

9110 Change in Taxes Payable

9200 Extraordinary Gains (Losses)

Answer to Exercise 2-9: (Dates have been omitted for simplicity)

Journal

1.	Cash	500	
	Notes Payable		500
2.	Feeder Livestock	500	
	Cash		500
3.	(None)		
4.	Cash	1500	
	Cash Sales of Market Livestock and Poultry	1500	
5.	Purchased Feed	100	
	Cash		100

LEDGER

Cash

Date	Description	Debits	Credits	Balance
Beginning Balance				*0*
1.	Borrowed money from bank	500		500
2.	Purchased steer		500	0
4.	Sold steer	1,500		1,500
5.	Paid bill		100	1,400

Feed Inventory Purchased for Use

Date	Description	Debits	Credits	Balance
Beginning Balance				*30*

Notes Payable

Date	Description	Debits	Credits	Balance
Beginning Balance				*0*
1.	Borrowed money from bank		500	500

Cash Sales of Market Livestock and Poultry

Date	Description	Debits	Credits	Balance
Beginning Balance				*0*
4.	Sold steer		1,500	1,500

Feeder Livestock

Date	Description	Debits	Credits	Balance
Beginning Balance				*0*
2.	Purchased steer	500		500

Purchased Feed

Date	Description	Debits	Credits	Balance
Beginning Balance				*0*
5.	Purchased feed	100		100

The Farmers' Ledger

1000 Cash

Date	Description	JE#	Debits	Credits	Balance
Beginning Balance					0
Jan. 2	Set up farm bank account	(1)	20,000		20,000
Feb. 1	Income tax payment	(31)		2,160	17,840
Feb. 12	Transfer from personal checking account	(3)	100		17,940
Mar. 1	Money borrowed to purchase land	(6)	120,000		137,940
Mar. 1	Money borrowed to purchase tractor	(7)	50,000		187,940
Mar. 1	Purchase of land	(9)		120,000	67,940
Mar. 1	Purchase of tractor	(11)		46,000	21,940
Mar. 10	Sale of bull to neighbor	(10)	900		22,840
Mar. 17	Transfer to personal checking account	(4)		150	22,690
Apr. 1	Employee paycheck	(26)		1,018	21,672
Apr. 10	FICA taxes and FIT paid to IRS	(27)		274	21,398
Apr. 10	FUTA tax paid to IRS	(28)		9.60	21,388.40
Apr. 10	SUTA tax paid to state agency	(29)		27.60	21,360.80
Apr. 17	Purchase of cattle tags	(22)		75	21,285.80
May 10	Purchase of feed	(20)		1,000	20,285.80
May 16	Purchase of feeder pigs	(15)		1,500	18,785.80
May 20	Investment in apple orchard	(14)		45,000	(26,214.20)
June 1	Payment for herbicide	(23)		500	(26,714.20)
July 10	Payment of real estate and property taxes	(30)		1,300	(28,014.20)
July 17	Sale of half of the feeder pigs	(16)	900		(27,114.20)
Aug. 1	Payment of insurance premium for next 12 months	(25)		1,200	(28,314.20)
Aug. 1	Lease payment on harvester	(13)		23,982	(52,296.20)
Aug. 16	Sale of grain at harvest time	(17)	12,000		(40,296.20)
Oct. 1	To record interest and principal payment	(8)		11,200	(51,496.20)
Oct. 31	Sale of grown hay to neighbor	(21)	600		(50,896.20)
Nov. 20	Sale of culled cows	(19)	1,250		(49,646.20)
Dec. 1	Truck expense to haul feeder cattle to market	(24)		150	(49,796.20)
Dec. 1	Sale of feeder calves	(18)	50,000		203.80

1500 Breeding Livestock

Date	Description	JE#	Debits	Credits	Balance
Beginning Balance					*0*
Jan. 2	Set up farm account for breeding cattle	(1)	76,000		76,000
Mar. 10	Sale of bull to neighbor	(10)		1,000	75,000
Nov. 20	Sale of culled cows	(19)		1,500	73,500

1600 Machinery and Equipment

Date	Description	JE#	Debits	Credits	Balance
Beginning Balance					*0*
Jan. 2	Transfer of personal assets (pickup truck and old tractor)	(1)	20,000		20,000
Mar. 1	Purchase of new tractor	(11)	50,000		70,000
Mar. 1	Trade-in of old tractor	(11)		5,000	65,000

1650 Office Furniture/Equipment

Date	Description	JE#	Debits	Credits	Balance
Beginning Balance					*0*
Jan. 2	Transfer of personal assets	(1)	1,000		1,000

1700 Perennial Crops and Natural Resources

Date	Description	JE#	Debits	Credits	Balance
Beginning Balance					*0*
May 20	Investment in apple orchard	(14)	45,000		45,000

1800 Land, Buildings' and Improvements

Date	Description	JE#	Debits	Credits	Balance
Beginning Balance					*0*
Jan. 2	Inherited farm land, buildings, and improvements	(5)	350,000		350,000
Mar. 1	Purchase of land	(9)	120,000		470,000

1910 Leased Assets

Date	Description	JE#	Debits	Credits	Balance
Beginning Balance					*0*
Aug. 1	Leased harvester	(12)	100,000		100,000

1980 Accumulated Depreciation

Date	Description	JE#	Debits	Credits	Balance
Beginning Balance					*0*
Jan. 2	To set up account	(1)		300	300
May 10	Sale of bull	(10)	300		0

2100 Taxes Payable

Date	Description	JE#	Debits	Credits	Balance
Beginning Balance					*0*
Jan. 2	To set up farm taxes payable	(2)		2,160	2,160
Apr. 1	To record payroll withholdings	(26)		182	2,342
Apr. 10	Payment of payroll withholdings	(27)	182		2,160

2310 Notes Payable Due within One Year

Date	Description	JE#	Debits	Credits	Balance
Beginning Balance					*0*
Jan. 2	To set up farm notes payable	(2)		10,000	10,000
Mar. 1	Money borrowed to purchase tractor	(7)		50,000	60,000
Oct. 1	To record annual principal payment	(8)	10,000		50,000

2400 Real Estate Notes Payable—Non-Current

Date	Description	JE#	Debits	Credits	Balance
Beginning Balance					*0*
Mar. 1	Money borrowed to purchase land	(6)		120,000	120,000

2600 Obligations on Leased Assets

Date	Description	JE#	Debits	Credits	Balance
Beginning Balance					*0*
Aug. 1	Leased harvester	(12)		100,000	100,000
Aug. 1	Lease payment	(13)	23,982		76,018

3100 Retained Capital

Date	Description	JE#	Debits	Credits	Balance
Beginning Balance					*0*
Jan. 2	To set up farm equity	(1)		116,700	116,700
Jan. 2	To set up farm equity	(2)	12,160		104,540

3110 Owner Withdrawals

Date	Description	JE#	Debits	Credits	Balance
Beginning Balance					0
Mar. 17	Transfer to personal checking account	(4)	150		150

3120 Non-Farm Income

Date	Description	JE#	Debits	Credits	Balance
Beginning Balance					0
Feb. 12	Transfer from personal checking account	(3)		100	100

3130 Other Capital Contributions/Gifts/Inheritances

Date	Description	JE#	Debits	Credits	Balance
Beginning Balance					0
Jan. 2	Inherited farmland, buildings, and improvements	(3)		350,000	350,000

4000 Cash Crop Sales

Date	Description	JE#	Debits	Credits	Balance
Beginning Balance					0
Aug. 16	Sale of grain at harvest time	(17)		12,000	12,000
Oct. 31	Sale of grown hay to neighbor	(21)		600	12,600

4100 Cash Sales of Market Livestock and Poultry

Date	Description	JE#	Debits	Credits	Balance
Beginning Balance					0
July 17	Sale of half of the feeder pigs	(16)		900	900
Dec. 1	Sale of feeder calves	(18)		50,000	50,900

4500 Gains (Losses) from Sale of Culled Breeding Livestock

Date	Description	JE#	Debits	Credits	Balance
Beginning Balance					0
Nov. 20	Loss on sale of culled cows	(19)	250		(250)

5000 Feeder Livestock

Date	Description	JE#	Debits	Credits	Balance
Beginning Balance					0
May 16	Purchase of feeder pigs	(15)	1,500		1,500

5020 Purchased Feed

Date	Description	JE#	Debits	Credits	Balance
Beginning Balance					0
May 10	Purchase of feed	(20)	1,000		1,000

6100 Wages Expense

Date	Description	JE#	Debits	Credits	Balance
Beginning Balance					0
Apr. 1	Wages for hired help	(26)	1,200		1,200

6110 Payroll Tax Expense

Date	Description	JE#	Debits	Credits	Balance
Beginning Balance					0
Apr. 10	FICA taxes and employee FIT paid to IRS	(27)	92		92
Apr. 10	FUTA tax paid to IRS	(28)	9.60		101.60
Apr. 10	SUTA tax paid to state agency	(29)	27.60		129.20

6310 Truck and Machinery Hire

Date	Description	JE#	Debits	Credits	Balance
Beginning Balance					0
Dec. 1	Truck to haul feeder cattle to market	(24)	150		150

6520 Herbicides, Pesticides

Date	Description	JE#	Debits	Credits	Balance
Beginning Balance					0
June 1	Purchase of pesticides	(23)	500		500

6630 Livestock Supplies, Tools, and Equipment

Date	Description	JE#	Debits	Credits	Balance
Beginning Balance					0
Apr. 17	Purchase of cattle tags	(22)	75		75

6700 Insurance

Date	Description	JE#	Debits	Credits	Balance
Beginning Balance					0
Aug. 1	Payment of insurance premium for next 12 months	(25)	1,200		1,200

6710 Real Estate and Personal Property Taxes

Date	Description	JE#	Debits	Credits	Balance
Beginning Balance					0
July 1	Payment of real estate and property taxes	(30)	1,300		1,300

8100 Interest Expense

Date	Description	JE#	Debits	Credits	Balance
Beginning Balance					0
Oct. 1	To record interest payment	(8)	1,200		1,200

8200 Gains (Losses) on Sales of Farm Capital Assets

Date	Description	JE#	Debits	Credits	Balance
Beginning Balance					0
Mar. 1	Loss on trade-in of old tractor	(11)	1,000		(1,000)
Mar. 10	Gain on sale of bull to neighbor	(10)		200	(800)

9100 Income Tax Expense

Date	Description	JE#	Debits	Credits	Balance
Beginning Balance					0
Feb. 1	Income tax payment	(31)	2,160		2,160

Unadjusted Trial Balance
Steve and Chris Farmer
December 31, 20X1

Accounts	Debits	Credits
1000 Cash	$ 203.80	$
1500 Breeding Livestock	73,500.00	
1600 Machinery and Equipment	65,000.00	
1650 Office Furniture and Equipment	1,000.00	
1700 Perennial Crops and Natural Resources	45,000.00	
1800 Land, Buildings and Improvements	470,000.00	
1910 Leased Assets	100,000.00	
2100 Taxes Payable		2,160.00
2310 Notes Payable Due within One Year		50,000.00

Accounts	Debits	Credits
2400 Real Notes Payable—Non-Current		120,000.00
2600 Obligations on Leased Assets		76,018.00
3100 Retained Capital		104,540.00
3110 Owner Withdrawals	150.00	
3120 Non-Farm Income		100.00
3130 Other Capital Contributions/Gifts/Inheritances		350,000.00
4000 Cash Crop Sales		12,600.00
4100 Cash Sales of Market Livestock and Poultry		50,900.00
4500 Gains (Losses) from Sale of Culled Breeding Livestock	250.00	
5000 Feeder Livestock	1,500.00	
5020 Purchased Feed	1,000.00	
6100 Wages Expense	1,200.00	
6110 Payroll Tax Expense	129.20	
6310 Truck and Machinery Hire	150.00	
6520 Herbicides, Pesticides	500.00	
6630 Livestock Supplies, Tools, and Equipment	75.00	
6700 Insurance	1,200.00	
6710 Real Estate and Personal Property Taxes	1,300.00	
8100 Interest Expense	1,200.00	
8200 Gains (Losses) on Sales of Farm Capital Assets	800.00	
9100 Income Tax Expense	2,160.00	
Totals	$766,318.00	$766,318.00

APPENDIX F

The ledger below shows the content of each of the accounts used by the Farmers in the journal entries discussed in Chapter 3 and the adjusting journal entries from Chapter 4, that is, journal entries (1) through (46), except entry (36). Many other expenses and revenues occur in a "real" farm business, but are excluded in this book to avoid repetition. (For example, the expense for livestock supplies for an entire year is likely to be considerably greater than $75.) The purpose of these illustrations and examples is to demonstrate the procedures for posting entries and preparing a trial balance. The ledger accounts include the number of each journal entry as references. Page references are normally used as cross-references in accounting software packages and should also be used in manual accounting systems.

GENERAL LEDGER
Steve and Chris Farmer

1000 Cash

Date	Description	JE#	Debits	Credits	Balance
Beginning Balance					*0*
Jan. 2	Set up farm bank account	(1)	20,000		20,000
Feb. 1	Income tax payment	(31)		2,160	17,840
Feb. 12	Transfer from personal checking account	(3)	100		17,940
Mar. 1	Money borrowed to purchase land	(6)	120,000		137,940
Mar. 1	Money borrowed to purchase tractor	(7)	50,000		187,940
Mar. 1	Purchase of land	(9)		120,000	67,940
Mar. 1	Purchase of tractor	(11)		46,000	21,940
Mar. 10	Sale of bull to neighbor	(10)	900		22,840
Mar. 17	Transfer to personal checking account	(4)		150	22,690
Apr. 1	Employee paycheck	(26)		1,018	21,672
Apr. 10	FICA taxes and FIT paid to IRS	(27)		274	21,398
Apr. 10	FUTA tax paid to IRS	(28)		9.60	21,388.40
Apr. 10	SUTA tax paid to state agency	(29)		27.60	21,360.80
Apr. 17	Purchase of cattle tags	(22)		75	21,285.80
May 10	Purchase of feed	(20)		1,000	20,285.80

May 16	Purchase of feeder pigs	(15)		1,500	18,785.80
May 20	Investment in apple orchard	(14)		45,000	(26,214.20)
June 1	Payment for herbicide	(23)		500	(26,714.20)
July 1	Payment of real estate and property taxes	(30)		1,300	(28,014.20)
July 17	Sale of half of the feeder pigs	(16)	900		(27,114.20)
Aug. 1	Payment of insurance premium for next 12 months	(25)		1,200	(28,314.20)
Aug. 1	Lease payment on harvester	(13)		23,982	(52,296.20)
Aug. 16	Sale of grain at harvest time	(17)	12,000		(40,296.20)
Oct. 1	To record interest and principal payment	(8)		11,200	(51,496.20)
Oct. 31	Sale of grown hay to neighbor	(21)	600		(50,896.20)
Nov. 20	Sale of culled cows	(19)	1,250		(49,646.20)
Dec. 1	Truck expense to haul feeder cattle to market	(24)		150	(49,796.20)
Dec. 1	Sale of feeder cattle	(18)	50,000		203.80

1100 Accounts Receivable

Date	Description	JE#	Debits	Credits	Balance
Beginning Balance					*0*
Dec. 31	Adjusting entry to record change in accounts receivable	(42)	3,500		3,500

1212 Feeder Livestock Inventory Purchased for Resale

Date	Description	JE#	Debits	Credits	Balance
Beginning Balance					*0*
Dec. 31	Adjusting entry for market value of purchased feeder pigs	(35)	750		750

1223 Feed Inventory Raised for Use

Date	Description	JE#	Debits	Credits	Balance
Beginning Balance					*0*
Dec. 31	Adjusting entry for market value of hay on hand	(33)	2,300		2,300

1224 Feed Inventory Purchased for Use

Date	Description	JE#	Debits	Credits	Balance
Beginning Balance					*0*
Dec. 31	Adjusting entry for value of purchased feed on hand	(32)	230		230

1231 Crop Inventory Raised for Sale

Date	Description	JE#	Debits	Credits	Balance
Beginning Balance					0
Dec. 31	Adjusting entry for market value of crop on hand	(34)	6,000		6,000

1300 Prepaid Expenses

Date	Description	JE#	Debits	Credits	Balance
Beginning Balance					0
Dec. 31	Adjusting entry for insurance paid in advance	(37)	700		700

1400 Cash Investment in Growing Crop

Date	Description	JE#	Debits	Credits	Balance
Beginning Balance					0
Dec. 31	Adjusting entry for expenditures in orchard	(38)	2,500		2,500

1500 Breeding Livestock

Date	Description	JE#	Debits	Credits	Balance
Beginning Balance					0
Jan. 2	Set up farm account for breeding cattle	(1)	76,000		76,000
Mar. 10	Sale of bull to neighbor	(10)		1,000	75,000
Nov. 20	Sale of culled cows	(19)		1,500	73,500
Dec. 31	Adjustment for change in quantity of raised breeding cows	(43)	6,000		79,500

1600 Machinery and Equipment

Date	Description	JE#	Debits	Credits	Balance
Beginning Balance					0
Jan. 2	Transfer of personal assets (pickup truck and old tractor)	(1)	20,000		20,000
Mar. 1	Purchase of new tractor	(11)	50,000		70,000
Mar. 1	Trade-in of old tractor	(11)		5,000	65,000

1650 Office Furniture and Equipment

Date	Description	JE#	Debits	Credits	Balance
Beginning Balance					0
Jan. 2	Transfer of personal assets	(1)	1,000		1,000

6110 Payroll Tax Expense

Date	Description	JE#	Debits	Credits	Balance
Beginning Balance					0
Apr. 10	FICA taxes and employee FIT paid to IRS	(27)	92		92
Apr. 10	FUTA tax paid to IRS	(28)	9.60		101.60
Apr. 10	SUTA tax paid to state agency	(29)	27.60		129.20

6310 Truck and Machinery Hire

Date	Description	JE#	Debits	Credits	Balance
Beginning Balance					0
Dec. 1	Truck to haul feeder cattle to market	(24)	150		150

6520 Herbicides, Pesticides

Date	Description	JE#	Debits	Credits	Balance
Beginning Balance					0
June 1	Purchase of pesticides	(23)	500		500

6630 Livestock Supplies, Tools, and Equipment

Date	Description	JE#	Debits	Credits	Balance
Beginning Balance					0
Apr. 17	Purchase of cattle tags	(22)	75		75

6700 Insurance

Date	Description	JE#	Debits	Credits	Balance
Beginning Balance					0
Aug. 1	Payment of insurance premium for next 12 months	(25)	1,200		1,200

6710 Real Estate and Personal Property Taxes

Date	Description	JE#	Debits	Credits	Balance
Beginning Balance					0
July 1	Payment of real estate and property taxes	(30)	1,300		1,300

6780 Depreciation Expense

Date	Description	JE#	Debits	Credits	Balance
Beginning Balance					0
Dec. 31	To record annual depreciation on non-current assets	(39)	22,150		22,150

6810 Change in Accounts Payable

Date	Description	JE#	Debits	Credits	Balance
Beginning Balance					0
Dec. 31	Adjusting entry to record change in accounts payable	(41)	340		340

6820 Change in Prepaid Insurance

Date	Description	JE#	Debits	Credits	Balance
Beginning Balance					0
Dec. 31	Adjusting entry for insurance paid in advance	(37)		700	(700)

6830 Change in Investment in Growing Crop

Date	Description	JE#	Debits	Credits	Balance
Beginning Balance					0
Dec. 31	Adjusting entry for expenditures in orchard	(38)		2,500	2,500

8100 Interest Expense

Date	Description	JE#	Debits	Credits	Balance
Beginning Balance					0
Oct. 1	To record interest payment	(8)	1,200		1,200

8110 Change in Interest Payable

Date	Description	JE#	Debits	Credits	Balance
Beginning Balance					0
Dec. 31	Adjusting entry to record change in interest payable	(40)	5,000		5,000

8200 Gains (Losses) on Sales of Farm Capital Assets

Date	Description	JE#	Debits	Credits	Balance
Beginning Balance					0
Mar. 1	Loss on trade-in of old tractor	(11)	1,000		(1,000)
Mar. 10	Gain on sale of bull to neighbor	(10)		200	(800)

9100 Income Tax Expense

Date	Description	JE#	Debits	Credits	Balance
Beginning Balance					0
Feb. 1	Income tax payment	(31)	2,160		2,160

9110 Change in Taxes Payable

Date	Description	JE#	Debits	Credits	Balance
Beginning Balance					*0*
Dec. 31	Adjusting entry for change in taxes owed	(44)	870		870
Dec. 31	Adjusting for current deferred taxes	(45)	1,450		2,320
Dec. 31	Adjusting for non-current deferred taxes	(46)	3,300		5,620

ADJUSTED TRIAL BALANCE
Steve and Chris Farmer
December 31, 20X1

Accounts	Debits	Credits
1000 Cash	$ 203.80	
1100 Accounts Receivable	3,500.00	
1212 Feeder Livestock Inventory Purchased for Resale	750.00	
1223 Feed Inventory Raised for Use	2,300.00	
1224 Feed Inventory Purchased for Use	230.00	
1231 Crop Inventory Raised for Sale	6,000.00	
1300 Prepaid Expenses	700.00	
1400 Cash Investment in Growing Crops	2,500.00	
1500 Breeding Livestock	79,500.00	
1600 Machinery and Equipment	65,000.00	
1650 Office Furniture and Equipment	1,000.00	
1700 Perennial Crops	45,000.00	
1800 Land, Buildings and Improvements	470,000.00	
1910 Leased Assets	100,000.00	
1980 Accumulated Depreciation		22,150.00
2000 Accounts Payable		340.00
2100 Taxes Payable		3,030.00
2200 Interest Payable		5,000.00
2310 Notes Payable Due within One Year		50,000.00
2400 Real Estate Notes Payable—Non-Current		120,000.00
2500 Non-Current Deferred Taxes		3,300.00
2510 Current Deferred Taxes		1,450.00
2600 Obligations on Leased Assets		76,018.00
3100 Retained Capital		104,540.00
3110 Owner Withdrawals	150.00	
3120 Non-Farm Income		100.00

Accounts	Debits	Credits
3130 Other Capital Contributions/Gifts/Inheritances		350,000.00
4000 Cash Crop Sales		12,600.00
4010 Changes in Crop Inventories		8,300.00
4100 Cash Sales of Market Livestock and Poultry		50,900.00
4500 Gains (Losses) from Sale of Culled Breeding Livestock	250.00	
4600 Change in Value Due to Change in Quantity of Raised Breeding Livestock		6,000.00
4700 Change in Accounts Receivable		3,500.00
5000 Feeder Livestock	1,500.00	
5010 Changes in Purchased Feeder Livestock Inventories		750.00
5020 Purchased Feed	1,000.00	
5030 Changes in Purchased Feed Inventories		230.00
6100 Wages Expense	1,200.00	
6110 Payroll Tax Expense	129.20	
6310 Truck and Machinery Hire	150.00	
6520 Herbicides, Pesticides	500.00	
6630 Livestock Supplies, Tools, and Equipment	75.00	
6700 Insurance	1,200.00	
6710 Real Estate and Personal Property Taxes	1,300.00	
6780 Depreciation Expense	22,150.00	
6810 Change in Accounts Payable	340.00	
6820 Change in Prepaid Insurance		700.00
6830 Change in Investment in Growing Crops		2,500.00
8100 Interest Expense	1,200.00	
8110 Change in Interest Payable	5,000.00	
8200 Gains (Losses) on Sales of Farm Capital Assets	800.00	
9100 Income Tax Expense	2,160.00	
9110 Change in Taxes Payable	5,620.00	
Totals	$821,408.00	$821,408.00

Farmers' Ledger Accounts after Posting Closing Entries

3110 Owner Withdrawals

Date	Description	JE#	Debits	Credits	Balance
Beginning Balance					0
Mar. 17	Transfer to personal checking account	(4)	150		150
Dec. 31	Closing entry	(48)		150	0

3120 Non-Farm Income

Date	Description	JE#	Debits	Credits	Balance
Beginning Balance					0
Feb. 12	Transfer from personal checking account	(3)		100	100
Dec. 31	Closing entry	(48)	100		0

3130 Other Capital Contributions/Gifts/Inheritances

Date	Description	JE#	Debits	Credits	Balance
Beginning Balance					0
Jan. 2	Inherited farmland, buildings, and improvements	(3)		350,000	350,000
Dec. 31	Closing entry	(48)	350,000		0

4000 Cash Crop Sales

Date	Description	JE#	Debits	Credits	Balance
Beginning Balance					0
Aug. 16	Sale of grain at harvest time	(17)		12,000	12,000
Oct. 31	Sale of grown hay to neighbor	(21)		600	12,600
Dec. 31	Closing entry	(47)	12,600		0

4010 Changes in Crop Inventories

Date	Description	JE#	Debits	Credits	Balance
Beginning Balance					*0*
Dec. 31	Adjusting entry for market value of hay on hand	(33)		2,300	2,300
Dec. 31	Adjusting entry for market value of crop on hand	(34)		6,000	8,300
Dec. 31	Closing entry	(47)	8,300		0

4100 Cash Sales of Market Livestock and Poultry

Date	Description	JE#	Debits	Credits	Balance
Beginning Balance					*0*
July 17	Sale of half of the feeder pigs	(16)		900	900
Dec. 1	Sale of feeder cattle	(18)		50,000	50,900
Dec. 31	Closing entry	(47)	50,900		0

4500 Gains (Losses) from Sale of Culled Breeding Livestock

Date	Description	JE#	Debits	Credits	Balance
Beginning Balance					*0*
Nov. 20	Loss on sale of culled cows	(19)	250		(250)
Dec. 31	Closing entry	(47)		250	0

4600 Change in Value Due to Change in Quantity of Raised Breeding Livestock

Date	Description	JE#	Debits	Credits	Balance
Beginning Balance					*0*
Dec. 31	Adjustment for change in quantity of raised breeding cows	(43)		6,000	6,000
Dec. 31	Closing entry	(47)	6,000		0

4700 Change in Accounts Receivable

Date	Description	JE#	Debits	Credits	Balance
Beginning Balance					*0*
Dec. 31	Adjusting entry to record change in accounts receivable	(42)		3,500	3,500
Dec. 31	Closing entry	(47)	3,500		0

5000 Feeder Livestock

Date	Description	JE#	Debits	Credits	Balance
Beginning Balance					0
May 16	Purchase of feeder pigs	(15)	1,500		1,500
Dec. 31	Closing entry	(47)		1,500	0

5010 Change in Purchased Feeder Livestock Inventories

Date	Description	JE#	Debits	Credits	Balance
Beginning Balance					0
Dec. 31	Adjusting entry for change in value of purchased feeder pigs	(35)		750	750
Dec. 31	Closing entry	(47)	750		0

5020 Purchased Feed

Date	Description	JE#	Debits	Credits	Balance
Beginning Balance					0
May 10	Purchase of feed	(20)	1,000		1,000
Dec. 31	Closing entry	(47)		1,000	0

5030 Change in Purchased Feed Inventories

Date	Description	JE#	Debits	Credits	Balance
Beginning Balance					0
Dec. 31	Adjusting entry for change in value of purchased feed	(32)		230	230
Dec. 31	Closing entry	(47)	230		0

6100 Wages Expense

Date	Description	JE#	Debits	Credits	Balance
Beginning Balance					0
Apr. 1	Wages for hired help	(26)	1,200		1,200
Dec. 31	Closing entry	(47)		1,200	0

6110 Payroll Tax Expense

Date	Description	JE#	Debits	Credits	Balance
Beginning Balance					0
Apr. 10	FICA taxes and employee FIT paid to IRS	(27)	92		92
Apr. 10	FUTA tax paid to IRS	(28)	9.60		101.60
Apr. 10	SUTA tax paid to state agency	(29)	27.60		129.20
Dec. 31	Closing entry	(47)		129.20	0

6310 Truck and Machinery Hire

Date	Description	JE#	Debits	Credits	Balance
Beginning Balance					0
Dec. 1	Truck to haul feeder cattle to market	(24)	150		150
Dec. 31	Closing entry	(47)		150	0

6520 Herbicides, Pesticides

Date	Description	JE#	Debits	Credits	Balance
Beginning Balance					0
June 1	Purchase of pesticides	(23)	500		500
Dec. 31	Closing entry	(47)		500	0

6630 Livestock Supplies, Tools, and Equipment

Date	Description	JE#	Debits	Credits	Balance
Beginning Balance					0
Apr. 17	Purchase of cattle tags	(22)	75		75
Dec. 31	Closing entry	(47)		75	0

6700 Insurance

Date	Description	JE#	Debits	Credits	Balance
Beginning Balance					0
Aug. 1	Payment of insurance premium for next 12 months	(25)	1,200		1,200
Dec. 31	Closing entry	(47)		1,200	0

6710 Real Estate and Personal Property Taxes

Date	Description	JE#	Debits	Credits	Balance
Beginning Balance					0
July 1	Payment of real estate and property taxes	(30)	1,300		1,300
Dec. 31	Closing entry	(47)		1,300	0

6780 Depreciation Expense

Date	Description	JE#	Debits	Credits	Balance
Beginning Balance					0
Dec. 31	To record annual depreciation on non-current assets	(39)	22,150		22,150
Dec. 31	Closing entry	(47)		22,150	0

6810 Change in Accounts Payable

Date	Description	JE#	Debits	Credits	Balance
Beginning Balance					0
Dec. 31	Adjusting entry to record change in accounts payable	(41)	340		340
Dec. 31	Closing entry	(47)		340	0

6820 Change in Prepaid Insurance

Date	Description	JE#	Debits	Credits	Balance
Beginning Balance					0
Dec. 31	Adjusting entry for insurance paid in advance	(37)		700	(700)
Dec. 31	Closing entry	(47)	700		0

6830 Change in Investment in Growing Crops

Date	Description	JE#	Debits	Credits	Balance
Beginning Balance					0
Dec. 31	Adjusting entry for expenditures in orchard	(38)		2,500	2,500
Dec. 31	Closing entry	(47)	2,500		0

8100 Interest Expense

Date	Description	JE#	Debits	Credits	Balance
Beginning Balance					*0*
Oct. 1	To record interest payment	(8)	1,200		1,200
Dec. 31	Closing entry	(47)		1,200	0

8110 Change in Interest Payable

Date	Description	JE#	Debits	Credits	Balance
Beginning Balance					*0*
Dec. 31	Adjusting entry to record change in interest payable	(40)	5,000		5,000
Dec. 31	Closing entry	(47)		5,000	0

8200 Gains (Losses) on Sales of Farm Capital Assets

Date	Description	JE#	Debits	Credits	Balance
Beginning Balance					*0*
Mar. 1	Loss on trade-in of old tractor	(11)	1,000		(1,000)
Mar. 10	Gain on sale of bull to neighbor	(10)		200	(800)
Dec. 31	Closing entry	(47)		800	0

9100 Income Tax Expense

Date	Description	JE#	Debits	Credits	Balance
Beginning Balance					*0*
Feb. 1	Income tax payment	(31)	2,160		2,160
Dec. 31	Closing entry	(47)		2,160	0

9110 Change in Taxes Payable

Date	Description	JE#	Debits	Credits	Balance
Beginning Balance					*0*
Dec. 31	Adjusting entry for change in taxes owed	(44)	870		870
Dec. 31	Adjusting for current deferred taxes	(45)	1,450		2,320
Dec. 31	Adjusting for non-current deferred taxes	(46)	3,300		5,620
Dec. 31	Closing entry	(47)		5,620	0

Long Version of Valuing Raised Breeding Livestock When Animals Are Sold or Have Died

In the long version, the following steps occur:

- The change in value due to age progression is calculated by determining the number of animals that were born or transferred into the next age category and then valuing that change.
- The number of head at the beginning of the year minus the number that were sold or died is the number that transferred to the next category.
- The number at the end of the year is calculated as the number at the beginning of the year minus the number sold and died minus the number transferred out (to the next age level) plus the number transferred in from the previous age level.
- The number born is the number transferred in to the youngest age category.
- The change in value from the previous category is the difference between the per head base value at the beginning of the year and the per head base value of the category that the animals transferred into.

These formulas are depicted as follows:

Number transferred in = Number transferred out from previous group

Number at beginning of year

− Number sold/died

− Number transferred out

+ Number transferred in

= Number at end of year

Beginning of year base value

− Base value of category transferred into

= Change in base value from previous category

- The change in value from previous category is calculated for each age category and multiplied by the number of head transferred into each age category to yield the change in value for each category.
- These values are then added to calculate the change in value due to age progression (excluding gains and losses from the sale of breeding livestock).

At the end of the second year, the number of Calves is two, the number of Bred Heifers is zero, and the number of Cows is ten. In the third year, eight new calves were born and the two calves that were born the previous year became bred heifers. The Farmers sold one of the bred heifers and two of the cows in the third year.

A valuation table indicating the changes in the third year is as follows:

Group Approach/Multiple Transfer Points/Long Version/Year 3/Base Values Unchanged:

Category	No. at end of 2nd year	No. sold	No. died	Transferred out	Born	Transferred in	No. at end of 3rd year	No. of new animals	Change in Base Value	Total Change
Calves	2	1	0	1	8	8	8	8	$400	$3,200
Bred Heifers	0	0	0	0		1	1	1	400	400
Cows	10	2	0			0	8	0	200	0
Total	12						17			$3,600

"Transferred out" refers to the number that transferred to the next age group. "Transferred in" refers to the number that transferred from the previous age group.

The following calculations are performed:

Calves: Number transferred out = Number at beginning of year
$$- \text{ Number sold/died}$$
$$= 2 - 1 = 1$$

Bred Heifers: Number transferred out = 0

Calves: Number transferred in = Number born = 8

Bred Heifers: Number transferred in = Number transferred out from
$$\text{calves} = 1$$

Cows: Number transferred in = Number transferred out from
$$\text{bred heifers} = 0$$

Calves: Number at end of year = Number at beginning of year
$$- \text{ Number sold/died} - \text{Number}$$
$$\text{transferred out} + \text{Number transferred in}$$
$$= 2 - 1 - 1 + 8 = 8$$

Bred Heifers: Number at end of year = $0 - 0 - 0 + 1 = 1$

Cows: Number at end of year = $10 - 2 - 0 + 0 = 8$

Calves: Change in Base Value = Base value of category transferred into
$$- \text{ beginning of year base value}$$
$$= \$400 - 0 = \$400$$

Bred Heifers: Change in value from previous category = $\$800 - 400 = \$400$

Cows: Change in value from previous category = $\$1,000 - 800 = \$200$

Change in value for categories = Number of new animals × Change in
$$\text{base value}$$

Calves:	$8 \times \$400 =$	$3,200
Bred Heifers:	$1 \times \$400 =$	400
Cows:	$0 \times \$200 =$	0

Change in value due to age progression: $3,600

The income statement reports the following:

Loss from Sale of Culled Breeding Livestock	(1,200)
Change in Value Due to Change in Quantity of Raised Breeding Livestock	3,600
Gross Revenue	$
Operating Expenses +/− Accrual Adjustments	$
Interest Expense +/− Accrual Adjustments	$
Net Farm Income from Operations	$
Gain Due to Changes in General Base Values of Breeding Livestock	$
Net Farm Income, Accrual Adjusted	$

The net effect of the loss on the sale of culled animals and the change in value due to age progression ($3,600 − 1,200) is the same net effect shown on the income statement prepared under the shortcut version ($1,200 + 1,200).

Partial Periods for Accelerated Depreciation Methods

If the Farmers decided to use the sum-of-years-digits method for the truck purchased on October 1, 20X1, depreciation for the first 12 months is $4,000. Only 3/12ths of $4,000 would be recorded in the year 20X1. A convenient way to allocate the first year's depreciation and all subsequent years' depreciation is to compute depreciation for one month at a time and construct a schedule. For the Farmers' truck, a monthly schedule would be constructed as follows:

October 20X1:	$4,000 ÷ 12 =	$ 333.33
November 20X1:	4,000 ÷ 12 =	333.33
December 20X1:	4,000 ÷ 12 =	333.33
Total depreciation for 20X1:		$ 999.99 (round up to $1,000.00)
January 20X2:	4,000 ÷ 12 =	333.33
February 20X2:	4,000 ÷ 12 =	333.33
March 20X2:	4,000 ÷ 12 =	333.33
April 20X2:	4,000 ÷ 12 =	333.33
May 20X2:	4,000 ÷ 12 =	333.33
June 20X2:	4,000 ÷ 12 =	333.33
July 20X2:	4,000 ÷ 12 =	333.33
August 20X2:	4,000 ÷ 12 =	333.33
September 20X2:	4,000 ÷ 12 =	333.33
October 20X2:	3,200 ÷ 12 =	266.67
November 20X2:	3,200 ÷ 12 =	266.67
December 20X2:	3,200 ÷ 12 =	266.67
Total depreciation for 20X2:		$3,799.98

The first 12-month amount of depreciation of $4,000 is allocated to the last three months in 20X1, beginning with the month in which the truck was purchased, and the first nine months of 20X2. In October, 20X2, the allocation of the second 12-month amount begins and continues through September 20X3. This allocation process proceeds until the full amount of the base has been allocated. The amount of depreciation recorded is the total for the months of January through December for each year. For 20X1, the depreciation expense is $1,000 and for 20X2 the depreciation expense is $3,799.98. The next 12-month amount of depreciation is $12,000 × 3/15 = $2,400.

A similar schedule can be set up if the Farmers decided to use the double-declining balance method:

October 20X1:	$6,000 ÷ 12 =	$ 500
November 20X1:	6,000 ÷ 12 =	500
December 20X1:	6,000 ÷ 12 =	500
Total depreciation for	20X1:	$1,500
January 20X2:	6,000 ÷ 12 =	500
February 20X2:	6,000 ÷ 12 =	500
March 20X2:	6,000 ÷ 12 =	500
April 20X2:	6,000 ÷ 12 =	500
May 20X2:	6,000 ÷ 12 =	500
June 20X2:	6,000 ÷ 12 =	500
July 20X2:	6,000 ÷ 12 =	500
August 20X2:	6,000 ÷ 12 =	500
September 20X2:	6,000 ÷ 12 =	500
October 20X2:	3,600 ÷ 12 =	300
November 20X2:	3,600 ÷ 12 =	300
December 20X2:	3,600 ÷ 12 =	300
Total depreciation for	20X2:	$5,400

As in the case for the sum-of-years-digits method, the first 12-month amount of $6,000 is allocated to the last three months of 20X1 and the first nine months of 20X2. The second 12-month amount is allocated to the last three months of 20X2 and the first three months of 20X3 and so on until the book value is equal to the salvage value. The next 12-month amount of depreciation expense is ($15,000 − 6,000 − 3,600) × 2 × (1 ÷ 5) = $2,160.

Reporting Income Taxes and Deferred Taxes

For many farm operations, income tax expense and deferred taxes will not necessarily show up on the farm financial statements. The approach for reporting these items depends on how the farm business has been organized. Sole proprietorships, partnerships, S-corporations, and limited liability companies are considered "pass-through" entities and the owners are responsible for the tax liabilities. The tax forms for some of these entities are schedules attached to the owner(s)' personal 1040 income tax forms. In these cases, the farm business is not liable for taxes but the owners are liable for the taxes on their businesses. C-corporations are separate legal entities from their owners and thus are liable for taxes. For these types of farm operations, the income tax expense and any deferred taxes are appropriately included in the farm financial statements.

Accelerated methods of depreciation: Methods that calculate a greater amount of depreciation (than the straight-line method would) during the early years of the asset's life and less in the later years of the asset's life, resulting in depreciation expense declining each year.

Accounting: An information system designed to provide financial information about an economic entity.

Accounting equation: A mathematical relationship between assets, liabilities, and equity.

Accounts: Records of the activities involving all financial statement items.

Accrual-adjusted approach: System in which certain adjustments are made to cash-basis financial statements at the end of the year to arrive at accrual numbers.

Accrual-adjusted financial statements: Financial statements prepared under the accrual-adjusted approach that include the year-end adjustments.

Accrual Adjusted Net Farm Income (NFI): The return, or profit, of the farm operation, reported on the income statement; also a measure of the return from the farm business to the farmer for unpaid labor and management.

Accrual Adjusted Net Income: Income before Taxes minus Income Tax Expense, similar to Net Income.

Accrual-basis system: System in which revenue is reported when it is earned, whether the cash has been received or not, and expenses are reported when they occur, not only when they are paid.

Accrued expenses: Expenses that have occurred for the farm business but have not yet been paid for.

Accrued revenues: Income that has been earned but cash has not yet been received.

Accumulated depreciation: The total amount of the cost of an asset that has been allocated for all the years since the asset was put into use.

Activity method of depreciation: Calculating depreciation expense based on the amount that the asset has been used during the period.

Adjunct account: A financial statement items that adds to another account.

Adjusted trial balance: Trial balance that includes adjusting journal entries.

Adjusting journal entries: Journal entries in which adjustments are recorded.

Allowance method: The balance sheet reports accounts receivable adjusted for an estimate of lost income from bad debts before notification occurs.

Amortization table: Table that lists the amounts of principal and interest that are paid with each cash payment.

Asset turnover ratio: Measure of the level of efficiency in using farm assets to generate farm revenue.

Assets: The items purchased by the farm business that are expected to earn money for the farm.

Automated accounting system: System of accounting involving the use of computer hardware and software, where the ledger is in a computer (not a book).

Bad debts or uncollectible accounts: Money owed to the farm business that might not be received.

Balance: Net amount of the increases and decreases in an account.

Balance Sheet: Financial statement that summarizes the financing and investing activities and the financial position of a farm business.

Bank charges: Charges against the farm bank account for service charges.

Bank credits: Collections or deposits made by the bank on behalf of the farm account.

Bank reconciliation schedule: Table explaining the differences between the bank statement and the ledger account.

Bank statement: A report of the cash transactions that occurred at the bank.

Base: The amount of the cost of the asset to be allocated during the life of the asset.

Base value method: Method of valuation that designates a base value for different categories of raised breeding livestock.

Bonds: IOUs issued by a company for money that the company has borrowed from the public.

Book value: Original cost minus the accumulated depreciation or the current market value. Also, the cost of an asset not yet allocated.

Capital leases: Special types of leases that, in effect, are installment purchases.

Capital replacement and term debt repayment margin: Measure of the farm producer's ability to generate sufficient funds for repayment of non-current debt and to replace capital assets. It also measures the ability to acquire capital assets or service additional debt.

Cash-basis system: System in which revenue is recorded only when cash is received and expenses are recorded only when cash is paid.

Cash flow projections: Indications of when and how much cash will flow into the farm business and when and how much will flow out to pay bills and service debts.

Cash surrender value: Amount of money that can be received from cashing in a whole life insurance policy.

Change in Accounts Payable: Adjustment item to reflect the amount of expense that has occurred but has not been paid.

Change in Accounts Receivable: Adjustment item to reflect the correct amount of revenue that has been earned.

Change in Crop Inventory: Adjustment item reflecting the change in revenue to the farm operation from crops that are raised.

Change in Excess of Market Value over Cost/Tax Basis of Farm Capital Assets: A record of the changes in market values of assets, resulting in increases (or decreases) in equity.

Change in Interest Payable: Adjustment item to report the correct amount of interest for the year.

Change in Investment in Growing Crops: Adjustment used so that the cost of growing perennial crops not yet harvested is reported as a prepaid expense.

Change in Market Livestock and Poultry Inventory: Adjustment item reflecting an increase in equity for the farm business from raising livestock.

Change in Non-Current Portion of Deferred Taxes: An equity item reported with Change in Excess of Market Value over Cost/Tax Basis of Farm Capital Assets to show the tax effects of the market value adjustments.

Change in Prepaid Insurance: Adjustment item used so that only the amount of insurance coverage used up is reported as the cost of insurance for the year.

Change in Purchased Feed Inventory: Adjustment item used to report the correct amount of purchased feed expense.

Change in Purchased Feeder Livestock Inventory: Adjustment item used to report the correct amount of purchased feeder livestock expense.

Change in Taxes Payable: Adjustment item used to report the correct amount of taxes for the year.

Change in Value Due to Change in Quantity of Raised Breeding Livestock: Adjustment for the change in the value of breeding stock due to age progression.

Chart of Accounts: List of all farm accounts of a farm operation.

Closing entries: The final journal entries prepared at the end of the year to prepare the accounts for the beginning of next year.

Closing the accounts: The process of transferring the balances of the income statement accounts and some of the equity accounts to the Retained Capital account.

Comparative financial statements: Financial statements with more than one year of data.

Contra account: A financial statement item that subtracts from another account.

Credit: The right-hand column of an account. When an entry is made on the right side, it is known as crediting the account.

Current assets: Assets that are either cash, will be sold for cash, or will be used up during the next year.

Current deferred taxes: Liabilities or assets related to future tax payments or deductions from accrual adjustments that will be resolved during the next year.

Current liabilities: Debts that are due within a year's time.

Current market value: The price obtained in the marketplace between a willing buyer and a willing seller.

Current market value method: Valuation method based on the amount of cash obtained by selling the asset in an orderly liquidation.

Current ratio: Measure of the ability of the farm business to pay short-term debts.

Debit: The left-hand column of an account. When an entry is made on the left side, it is known as debiting the account.

Debt to assets ratio: Measure of the percentage of assets financed by creditors.

Debt to equity ratio: Measure of the extent to which creditors have supplied capital relative to the amount of capital supplied by owners and the accumulated results of operations.

Declining-balance method: Calculating depreciation expense by multiplying the book value of an asset times a multiple of the straight-line rate.

Deferred tax liabilities: Differences in tax liabilities between the cash-basis income and accrual-adjusted income that will be resolved in future years when the cash transactions are completed.

Deposits in transit: Cash amounts that have been deposited but are absent from the bank statement because they have not yet cleared the bank.

Depreciation: Process of allocating the cost of non-current purchased assets.

Depreciation Expense: An expense related to the annual allocation of the cost of purchased assets.

Depreciation expense ratio: Measure of the percentage of gross revenue used for depreciation expense.

Development phase of perennial crops: The time from purchasing and planting perennial plants until the first year that the plants produce a crop.

Direct write-off method: The farm accountant adjusts accounts receivable only when certain that the money will not be received.

Disclosure by Notes (or Notes to the Financial Statements): Additional information reported with financial statements that includes schedules and explanations of computations.

Discounted cash flow method: Valuation method based on the estimated net cash flows from the use of the asset over the estimated useful life.

Double-entry accounting: Method of recording journal entries in which the accountant records at least one debit and at least one credit for every transaction.

Equity: The owner's interest in the business (that is, the value of the farm operation to the owner(s)). Also, the difference between the total assets and the total liabilities.

Equity to assets ratio: Measure of the percentage of assets financed by owners.

Errors: Mistakes made by either the bank or the farm accountant in recording cash transactions.

Estimated tax rate: An estimate of the percentage of income paid for income taxes.

Estimated useful life: Length of time in years that the asset will be used by the farm operation.

Expense accounts: Records of the costs of operating the farm for the year.

Expenses: The costs of operating the farm.

Extraordinary gains and losses: Changes in equity that result from transactions or occurrences that are infrequent and unusual.

Farm Financial Standards Council (FFSC): A group that has developed and published a set of recommendations in an attempt to standardize accounting procedures for agricultural operations.

Farm Financial Statements: Documents that report financial accounting information about an agricultural operation in specified formats.

Farm Revenues: Money earned by the farm business from the production or sale of farm products and gains from operating activities.

Financial accounting: Accounting systems that produce financial statements for managers, owners, and outside parties.

Financial Accounting Standards Board (FASB): Organization that has been developing financial accounting standards for companies in the United States.

Financial accounting systems: The guidelines for preparing cash-basis, accrual-basis, or accrual-adjusted financial statements.

Financial analysis: Set of procedures in which an economic entity's financial position and financial performance is evaluated.

Financial efficiency: The ability of the farm business to use its assets to generate revenues and the effectiveness of its operating, investing, and financing activities.

Financial Guidelines for Agricultural Producers (the Guidelines): The recommendations of the FFSC for financial accounting for agricultural producers.

Financial performance: Measures of how profitable the economic entity was over a specified period, usually a year.

Financial position: An economic entity's financial state on a specified date.

Financial statements: Reports submitted to outside parties that summarize the results of the financial activities of an economic entity.

Financial transaction: Activity or event that affects the financial position or financial performance of the company.

Financing activities: Activities in which capital is provided to the farm business.

First-In-First-Out (FIFO): Assumption used in valuing inventory in which the most recently occurring purchase price is the one used to value the items on hand so that the inventory that remains is valued according to the most recent purchase price.

Forward contract: Contract in which the producer is obligated to make delivery of crops or livestock.

Fraction of the year: The amount of time covered by a loan period less than one year.

Full cost absorption method: Keeping accurate records of all costs associated with raising breeding livestock.

Gains: Financial benefits from various activities. Increases in equity.

Gains and Losses on Sales of Farm Capital Assets: Gains and losses from the sale or trade-in of assets other than culled breeding livestock; the difference between cash received from the sale or trade-in and the book value of the asset sold.

Gains/Losses Due to Changes in General Base Values of Breeding Livestock: An adjustment to reflect the change in the value of raised breeding livestock due to the change in the base values.

Generally Accepted Accounting Principles (GAAP): Rules and guidelines that provide a standardized system of financial accounting.

Government loan programs: Program in which a farm producer is allowed to offer crops as collateral for a loan with the Commodity Credit Corporation (CCC) at a specified loan rate (principal) for up to nine months.

Gross Revenue: The sum of the revenue items.

Group value approach: The number of head in each age category are counted and multiplied by the base value for that group.

Historical cost: The original cost or purchase price of the asset.

Historical cost method: Valuation method based on the amount paid to acquire the asset.

Historical cost principle: Reporting assets at book value.

Horizontal analysis: Using percentage changes from one year to the next for each of the items on the financial statements to analyze the performance of a farm business.

Income before Taxes: A subtotal on the income statement calculated by adding or subtracting Other Revenue and Other Expenses from Accrual Adjusted Net Farm Income.

Income Statement: Financial statement that reports on the financial performance of a farm business.

Income Tax Expense: The amount paid for income taxes, plus certain adjustments.

Individual animal approach: Maintaining valuation records for each animal.

Installment loans: Loans that require periodic payments, which consist of interest and a certain amount of principal.

Intangible assets: Assets that are nonphysical in nature.

Interest Expense: The amount of money paid for interest on loans from a lender.

Interest expense ratio: Measure of the percentage of gross revenue used for interest expense.

Interest rate: Usually an annual percentage of the principal that would be paid in the form of interest after one year.

Inventory: An asset (purchased or raised) that will be used up or sold within a year as part of the normal operating activities of the farm business.

Investing activities: Activities involved in using money to buy and sell items that the farm needs to produce farm products.

Investments (on the Statement of Owner Equity): Any additions of cash or other personal assets that the owner (or others, such as family members) contributed to the farm business since the beginning of the year.

Investments by owners: The owners' contributions of their own cash or other assets to the farm business.

Journal: The book in which transactions are initially recorded.

Journal entry: Each recording in a journal, consisting of recording the date of the transaction, the accounts involved, the debits and credits, and the amount of money or value involved in the transaction.

Last-In-First-Out (LIFO): Assumption used in valuing inventory in which the most recently occurring purchase price is the one used to value the sold items so that the inventory that remains is valued according to the oldest purchase price.

Ledger: Collection of all of the accounts for a farm operation.

Lessee: The party that is using an asset owned by another party.

Liabilities: Types and amounts of money owed to outside parties.

Liquidity: The ability of a farm business to meet its financial obligations when due.

Losses: The opposite of gains. Decreases in equity.

Lump-sum purchase: A purchase in which more than one asset is purchased as a package for a single purchase price.

Manual accounting system: System in which all accounting is done by hand without the use of a computer and computer software, where the ledger is a book in which each account is listed on a separate page and all of the transactions affecting that account are recorded there.

Market-based financial statements: Financial statements that report the market values for assets.

Market value: The price that a buyer and a seller can agree upon in an exchange of an asset.

Measurement: Performing calculations for financial statement items.

Net Farm Income: Income number used to evaluate the financial performance of the farm business.

Net Farm Income from Operations: A version of Operating Income that includes some adjustments for operating expenses and interest expense.

Net farm income from operations ratio: Measure of the percentage of profit from farm operations relative to gross revenue.

Net income or Net profit: The "bottom line" on the Income Statement and the result of the difference between the revenues and the expenses.

Net of tax: The amount shown on the income statement that indicates the amount of the gain or loss after the tax effect has been subtracted.

Net realizable value (NRV): The estimated sale price (the amount that the inventory could be sold for) less estimated costs of disposal.

Net realizable value method: Valuation method based on the amount that the asset could be sold for, less any direct selling costs.

Non-Current assets: Assets used in the operation of the farm business that are expected to last more than one year.

Non-Current Deferred Taxes: Liabilities or assets related to future tax payments or deductions from accrual adjustments that will not likely be resolved during the next year.

Non-Current liabilities: Debts that will take longer than one year to pay off.

Non-Farm Income: Any money from other jobs or businesses that the owner earned and contributed to the farm business during the current year.

Normal balance: The debit or credit where increases in an account are recorded.

Operating activities: The day-to-day activities of producing or selling crops and livestock, managing hired help, and so on.

Operating expense ratio: Measure of the percentage of gross revenue used for operating expenses.

Operating Expenses: The amount paid for supplies and other costs that are required to operate the farm business, any losses from operating activities, and certain adjustments.

Operating Income: The difference between the revenues and the operating expenses.

Operating leases: Rental arrangements that do not meet the criteria for a capital lease.

Operating profit margin ratio: Measure of the return per dollar of gross revenue.

Orderly liquidation: An exchange in which neither the buyer or seller are forced to engage in the transaction.

Original cost: The purchase price of an asset.

Other Capital Contributions/Gifts/Inheritances: Contributions to the owner's equity by the owner, when a personal item is put to use in the farm business, or someone other than the farm owners, such as when the parents in a farm family transfer the ownership of land or other assets to children in the family as a part of an inheritance.

Other Expenses: Costs associated with financing and investing activities and farm income taxes.

Other Revenues: Money earned from other types of farm activities besides the production and sale of farm products and operating gains.

Outstanding checks: Checks written but not yet cleared by the bank.

Owner withdrawals: Cash from the farm bank account used to pay for personal expenses.

Payroll: Wages paid for hired labor and the associated taxes.

Permanent accounts: The balance sheet accounts that are not closed because they report the results of activities on a continuous basis.

Posting: Transferring the information from each journal entry to the accounts in the ledger.

Prepaid expenses: Items paid for in advance, such as supplies or other purchases that are used up over time, recorded as expenses when purchased.

Present value factor (PVF): A multiplier that discounts payments to their present value.

Present value of lease payments: The amount of future lease payments without the interest.

Principal: The amount of money that is borrowed.

Productive phase of perennial crops: The years that the perennial plants are producing a crop.

Profitability: The ability of the farm business to generate a profit.

Promissory note: A written promise to pay back a loan plus interest by a certain date.

Rate of return on farm assets: Measure of the ability of the farm operator to earn a reasonable profit (return) on the assets under the operator's control.

Rate of return on farm equity: Measure of the ability of the farm business to generate a profit on owner equity.

Relevance: The capacity of information to make a difference to parties that make decisions based on the information.

Reliability: Ability of the information to truly represent what it says it represents, verifiable by another party, and unbiased and free of errors.

Repayment capacity: The ability of the farm business to pay back loans from farm and non-farm income.

Retained Capital: The equity from the accumulated net income of the farm operation (retained earnings) and the net contributions as seen on the Statement of Owner Equity.

Revenue accounts: Records of the amount of money earned by the farm operation from the production or sale of farm products.

Revenue recognition: Concept that helps to explain when revenue has been earned.

Revenues: Money earned by an economic entity from the production or sale of products or services.

Salvage value: An estimate of the value of the asset at the end of its useful life.

Simple interest: Interest that is not compounded (added to the principal periodically before the loan is paid off).

Single-entry accounting: Method of recording journal entries that does not recognize both sides of a transaction. Asset accounts are always increased in the debit side and decreased on the credit side.

Solvency: The ability of the farm business to continue to operate as a viable business regardless of external forces that may occur and the ability of the farm business to pay all financial obligations if all assets were sold.

Source documents: Written evidence of a financial transaction.

Statement of Cash Flows: Financial statement that reports on the financing, investing, and operating activities involving cash.

Statement of Owner Equity: Financial statement that reports the financing activities involving the owner in more detail than the Balance Sheet and the accumulated profit or loss of the farm business.

Stocks: Certificates of ownership in a company sold to the public.

Straight-line method of depreciation: The method of allocating the cost of an asset evenly over the life of the asset.

Sum-of-years-digits method: Calculating depreciation expense by multiplying the base of an asset times a smaller fraction each year.

Supplementary schedules: Tables prepared for all loans and included in the notes to the financial statements, listing principal amounts, interest amounts, due dates, and other terms of the loan. Also prepared for other financial statement items, listing details of the items.

Tangible assets: Assets that are physical in nature.

Tax-based depreciation method: Method that calculates depreciation according to tax rules.

Tax basis or taxable basis: The value of an asset based on the amount of cash paid for the asset minus tax depreciation.

Taxable income: Income based on cash-basis income and tax laws that forms the basis for calculating the amount of income tax to pay.

Term debt and capital lease coverage ratio: Measure of the ability of the farm business to cover all term debt and capital lease payments.

Temporary accounts: The income statement and statement of owner's equity accounts that report the activities for one year only.

Transfer points: The movement of breeding animals through age categories during their normal life cycle.

Trend analysis: Calculating the percentage change of financial items from a base year.

Trial balance: List of the balances of all accounts in the ledger.

Unadjusted Trial Balance: Trial balance without the accounts used in adjusting journal entries.

Valuation Equity: The differences between cost and market values for non-current assets, including the tax effects of these differences.

Weighted average method: Computes an average cost of inventory taking into account the quantity that was purchased.

Withdrawals (or Owner Withdrawals): Any cash or other assets used by the owner for personal purposes.

Working capital: The amount of funds available after the sale of current farm assets and payment of all current farm liabilities.

(Date)   1900 Investments in Cooperatives and Other Investments XXX
            1000 Cash   XXX

Income is received in the form of dividends from investments in stock or interest from investments in bonds. Dividend income and interest income can be recorded as Miscellaneous Revenue.

(Date)   1000 Cash   XXX
            4800 Miscellaneous Revenue   XXX

- Some assets, such as land, might be purchased strictly for future resale. Land sells either as a whole parcel or in lots for real estate development.

The purchase is recorded just as the other investments indicated above. These investments might or might not generate any income. The intention may be to own the investment until the market price increases sufficiently to produce a significant gain on the sale. Land owned for this purpose could generate income if leased to another party. In that case, rental income can be recorded as Miscellaneous Revenue just like the investments indicated above.

- Some investments are stock purchases of other companies (less than 50 percent ownership) in which the investor can exert significant influence over the operating and financing policies of the purchased company.

The accounting for these kinds of investments is similar to investments in farm cooperatives.

Changes in market values of investments should be recorded for stocks and bonds that are intended to be sold within the next year, even if market-based financial statements are not prepared. Sometimes these investments are called **marketable securities** to differentiate them from the investments that are not intended to be sold in the near future. When financial statements are prepared, the following journal entries should be recorded if the market values at that time are higher than the original cost of the investment.

Dec. 31   1900 Investments in Cooperatives and Other Investments   XX
            3010 Change in Excess of Market Value over Cost   XX

The debits and credits are reversed if the market values at that time are lower than the original cost. The amount is the difference between the original cost and the current market value. The FFSC Guidelines do not recommend reporting market values for other investments, so changes in market values are not reported for them as they are for other assets in the farm business.

The farm accountant records the sale of any of the investments like other non-current farm assets. However, investments are not depreciated as non-current assets used in the farm operation, so there is no entry for accumulated depreciation. Gains or losses are recorded when the cash received from the sale is different from the balance in the investment account pertaining to the investment that was sold. For investments in stocks and bonds without significant influence in the management of the company, the balance includes the original cost and changes in market value for investments that will be sold within the next year. For investments in farm cooperatives and stock purchases where there is significant influence in the management of the company, the balance includes the original cost and adjustments for income and dividends. Clearly, if the farm operation has many investments, the farm accountant will have to keep separate accounting records for each investment to keep track of the adjustments.

Life Insurance Policies

Like other businesses, many farm operations will purchase life insurance policies on key people in the business and name the business as the beneficiary. Life insurance policies are a type of investment if the policy is a whole life policy. Whole life policies can be cashed in for money (called the **cash surrender value**) if the policyholder wishes to discontinue the coverage. Another popular type of life insurance policy is term life insurance, in which the policyholder pays the premiums to the life insurance company only until a certain age. Benefits can be received only if the insured person dies before that age. The policyholder cannot cash in a term life insurance policy if the policyholder decides to discontinue the policy. The cash surrender value of whole life policies increases each time that the policyholder pays the premium to the insurance company. The increase in cash value can be read from a report that the insurance company provides on the cash value of whole life policies. The policyholder records this increase whenever a life insurance premium is paid.

The original cost of the policy is not recorded as an investment. The first journal entry records the life insurance expense when making the payment. With each subsequent payment, the policyholder records the increase in the cash value along with the insurance expense. As a result, the investment account for the insurance policy is updated each year and is reported on the balance sheet at market value.

Suppose the Farmers pay a life insurance premium of $2,300 annually. They record the first payment as life insurance expense:

| (Date) | 6700 Insurance | 2,300 | |
| | 1000 Cash | | 2,300 |

The next year, the first report for the Farmers indicates that the policy has a cash value of $1,400. This increase in value offsets the insurance expense, because it represents an investment to the Farmers. The journal entry to record the payment of the insurance premium reflects these facts:

(Date	6700 Insurance	900	
	1900 Investments in Cooperatives and Other Investments	1,400	
	1000 Cash		2,300

Retirement Accounts

Retirement accounts begin with a purchase or deposit. Many types of retirement accounts exist, including savings accounts, traditional IRAs, and Roth IRAs, to name a few. Retirement accounts are considered personal assets and are not reported on farm financial statements unless combined statements are prepared. If combined statements are prepared, a debit to the personal asset account and a credit to cash are recorded at the time of the initial investment. Subsequent contributions to the fund are recorded in the same way.